EasyTerms™
Terminology Guidebook
for Cell Biology

Copyright 2010, Ed Creager

This edition of EasyTerms is one in a series of simple-to-use, college-level terminology guidebooks.

Although these guidebooks were originally intended for college students, many High School students will also find them helpful as they prepare for college.

Other topics covered in existing or forthcoming editions:

- Anatomy & Physiology (Human)
- Biology
- Biochemistry
- Botany
- Business Management
- Ecology

- Genetics
- Microbiology
- Nursing
- Nutrition
- Psychology
- Zoology

EasyTerms can help support your educational advancement and can boost the vocabulary of almost anyone who reads it.

For more information on these and other publications, please visit this site:

AplecreekBooks.weebly.com

and please note the author's "signature book" entitled,

"The Money-Saving Idea Book: Inside Tips for Starving, Students, Frugal Seniors and Every Financial Survivor."

("The Money-Saving Idea Book" © and ™, Ed Creager, 2009.)

Foreword

This Cell Biology edition is a simple-to-use, college-level* terminology guidebook and is part of the EasyTerms reference series. In the book, terms are arranged alphabetically within appropriate topic areas. The complete index makes it easy to find any term and its definition.

* These books can also help High School students prepare so that, before they attend college, they'll already know a considerable amount of the terminology they'll need.

A substantial number of the terms defined here have additional definitions outside the scope of the subject being covered. More general definitions and additional meanings, if sought, are to be found in less specialized publications such as dictionaries and encyclopedias.

Please check this website...

ApplecreekBooks.weebly.com

...for more information on other available books.

Many of the books offered are also available at major online retailers.

Important Notice:

The resources provided hereby, including websites, books and related materials, are intended to provide accurate information regarding the subject matter. All products and services are provided with the understanding that neither the author nor the publisher is engaged in rendering legal, accounting, or other professional or scholarly advice. If expert assistance is needed, the services of a competent professional should be obtained.

EasyTerms™
Terminology Guidebook

<u>Table of Contents</u>

The terms that follow are divided into the topics shown below. The page number on which the topic begins is given. Within each topic, the terms are arranged alphabetically.

Introduction	1	DNA Replication	66
Basic Chemistry	5	Cell Division	69
Carbohydrates	9	Genetics / Mutations	75
Lipids	13	Protein Synthesis	81
Proteins	16	Microbes and Microbial Genetics	87
Enzymes	20	Reproduction	91
Organelles	25	Development	96
Nuclear Structures	32	Evolution	100
Membranes	34	Kinds of Nutrition	102
Membrane Functions	39	Nerve Cells	104
Energy	45	Muscle Cells and Movement	111
Energy-Capturing Metabolism	50	Immunology	116
Photosynthesis	55	Microscopy / Other Techniques	121
Nucleic Acids	60	Index	

Introduction

1. **Angstrom**

   A unit of length equal to 0.1 nanometer.

2. **biochemistry**

   Study of molecules that make up living organisms.

3. **biology**

   The study of life.

4. **biosphere**

   An interconnected system over the earth's surface in which organisms exist.

5. **biotechnology**

   The use of a natural biological system to make a product or achieve a particular end.

6. **cell**

   A basic functional unit of a living organism.

7. **cell theory**

   A theory stating that living things are composed of cells.

8. **compartmentalization**

   Location of cellular functions with particular organelles.

9. **cytology**

   The study of cells.

10. **cytoplasm**

    Cell substance, excluding the nucleus.

11.   **data**

Observations from an experiment.

12.   **deductive reasoning**

Reasoning from a general statement to a specific case.

13.   **development**

Process of increasing in complexity.

14.   **environment**

Changeable surroundings around a cell or living organism.

15.   **eukaryotic**

Having a nucleus and membrane-bound organelles.

16.   **excitability**

Responsiveness to a stimulus.

17.   **feedback**

The use of the output of a process to influence the process.

18.   **genetics**

The study of mechanisms of heredity.

19.   **half-life**

Time for half a radioactive substance to degrade itself.

20.   **hypothesis**

A possible answer to a question; a possible explanation for observations that can be used to predict future outcomes.

21.   **in vitro**

In glass.

22.   in vivo

In life.

23.   inductive reasoning

Development of a general statement from a collection of observations.

24.   liter

The basic metric unit of fluid volume; 1.06 quart.

25.   lumen

Internal space or cavity formed by a membrane or cluster of cells.

26.   mass

The amount of matter in an object.

27.   median

Toward or in the middle.

28.   meter

The basic metric unit of length; 39.37 inches.

29.   nanometer

Unit of measure equal to one billionth of a meter.

30.   physiology

The study of life functions.

31.   prokaryotic

Lacking a nucleus and membrane-bound organelles.

32.   protoplasm

Cell substance; literally, first formed.

33. **resistivity**

Electrical resistance in biological membranes.

34. **responsiveness**

Ability to react to a stimulus.

35. **scientific method**

A method of collecting and testing data pertaining to a scientific question.

36. **secretion**

A cell product; the active transport of substances from the blood to the kidney filtrate.

37. **theory**

An explanation that accounts for many observations.

38. **tissue**

A group of similar cells, including intercellular substances, that carry out a particular function.

39. **weight**

The amount of gravitational force exerted on an object.

Basic Chemistry

40. acid

An ionizing substance that donates hydrogen ions.

41. alkaline

Basic, able to accept hydrogen ions.

42. anion

A negatively charge ion.

43. atom

Smallest particle that retains properties of an element.

44. atomic number

The number of protons in the nucleus of an atom.

45. atomic weight

The total number of protons and neutrons in an atom; the average number if there are isotopes of the element.

46. base

An ionizing substance that accepts hydrogen ions or reacts with an acid to form a salt.

47. buffer

A substance that resists pH change by holding or releasing hydrogen ions in a solution.

48. cation

A positively charged ion.

49. charge repulsion

Force separating like charges.

50.   colloid

Glue-like; a particle in a colloidal dispersion.

51.   colloidal dispersion

A state of matter with small particles suspended in a medium.

52.   compound

A substance with two or more elements combined in definite proportion.

53.   electron

A negatively charged particle that continually moves around the nucleus of an atom.

54.   element

A fundamental unit of matter.

55.   gram molecular weight

The quantity of a substance (in grams) equal to its molecular weight.

56.   hydrolysis

The splitting of a molecule with the addition of water.

57.   ion

A charged atom or group of atoms.

58.   isotope

An atom having a different number of neutrons than certain other atoms of the same element.

59.   kinetic

Pertaining to the energy of motion.

60.   mixture

Two or more substances combined in any proportions and retaining their individual properties.

61.   mole

A gram molecular weight.

62.   molecule

The smallest quantity of a substance that retains its chemical properties.

63.   neutron

An uncharged particle in the nucleus of an atom.

64.   nucleus

Central part of an atom or a cell.

65.   organic

Containing carbon.

66.   oxidation

Addition of oxygen or loss of electrons in a chemical reaction.

67.   pH

The negative logarithm of the hydrogen ion concentration; a scale for expressing acidity or alkalinity.

68.   potential energy

Energy due to position and capable of being released, as in a rock at the top of a hill.

69.   proton

A positively charged particle in the nucleus of an atom.

70.   radiation

Spreading from a center; the emission of electromagnetic particles and waves.

71.   reactant

A substance that enters into a chemical reaction.

72. **reduction**

Gain of an electron or loss of oxygen in a chemical reaction.

73. **solute**

A dissolved substance.

74. **solution**

A liquid containing dissolved substances.

75. **solvent**

A substance in which other substances can dissolve.

76. **sphere of hydration**

Aggregation of water molecules around an ion.

77. **trace element**

A chemical element normally present in very small amounts in living material.

78. **valence**

An ion's charge.

Carbohydrates

79. amylopectin

A form of starch molecule containing branched chains of glucose units.

80. amylose

A form of starch molecule consisting of straight chains of glucose units.

81. asymmetric carbon atom

A carbon atom with four different substituents.

82. biopolymer

Biological macromolecule that consists of many repeated subunits.

83. carbohydrate

An organic compound having several alcohol groups and an aldehyde or ketone group.

84. cellulose

Polysaccharide of glucose subunits found in plant cell walls.

85. chitin

Polysaccharide of N-acetylglucosamine units in insect and crustacean skeletons.

86. disaccharide

A molecule having two sugar (saccharide) units held together by a glycosidic bond.

87. Fischer projection

Model of a molecule that reflects its three-dimensional structure on a sheet of paper.

88. functional group

A component of a molecule that participates in a chemical reaction.

89.   **glucose**

Six-carbon aldehyde sugar that is the starting molecule in many metabolic pathways.

90.   **glycogen**

Branched polysaccharide consisting of glucose subunits and stored in animal cells.

91.   **Haworth projection**

Molecular model that depicts spatial relationships in parts of a molecule.

92.   **hemicellulose**

Polymers deposited with cellulose in cell walls of plants and fungi.

93.   **heptose**

Seven-carbon sugar.

94.   **hexose**

Six-carbon sugar.

95.   **homeopolymer**

Polymer having only one kind of subunit.

96.   **isomer**

A molecule having the same kinds and number of atoms as another molecule, but arranged differently.

97.   **ketosugar**

Sugar with a ketone as its functional group.

98.   **lactose**

Disaccharide with a glucose and a galactose unit.

99.   **lignin**

Cellulose polymer that strengthens cell walls of plants and fungi.

100. **macromolecule**

A very large molecule such as a protein or nucleic acid.

101. **maltose**

Disaccharide consisting of two glucose units.

102. **monomer**

Organic molecule that serves as subunit for a macromolecule.

103. **monosaccharide**

A simple sugar.

104. **mucopolysaccharide**

Chemical substance in the fuzzy layer of many animal cells.

105. **noncellulosic matrix**

Plant and fungal cell wall component containing hemicellulose, lignin, pectin, and extensin.

106. **pectin**

Polymer found with cellulose in cell walls of plants and fungi.

107. **pentose**

Five-carbon sugar.

108. **polymer**

A molecule consisting of repeating units.

109. **polysaccharide**

A molecule consisting of many saccharide units connected by glycosidic bonds.

110. **ribose**

Five-carbon sugar in RNA.

111. **starch**

Polysaccharide consisting of glucose units stored by plants.

112. **stereoisomer**

Compound having the same kind and number of atoms as another compound, but in a different spatial arrangement.

113. **storage macromolecule**

Polymer of one or more subunits stored by cells; starch or glycogen.

114. **sucrose**

Disaccharide comprised of glucose and fructose; table sugar.

115. **tetrose**

Four-carbon sugar.

116. **triose**

Three-carbon sugar.

Lipids

117. bile acid

Steroid that emulsifies fats in the intestine.

118. ceramide

Lipid containing sphingocine and a fatty acid.

119. cerebroside

Glycolipid containing ceramide and a monosaccharide.

120. cholesterol

Lipid in animal cell membranes and steroid precursor.

121. diglyceride

Diacylglycerol; glycerol and two fatty acids.

122. essential fatty acid

Fatty acid required in the diet because the body cannot make it.

123. fatty acid

A long hydrocarbon chain with a carboxyl group at one end.

124. glycerol

Three-carbon alcohol that typically combines with fatty acids.

125. glycolipid

A molecule that contains both carbohydrate and lipid components.

126. glycosphingolipid

Sphingolipid to which one or more sugars are attached.

127. hydrophilic

Attracted to water.

128. hydrophobic

Tending to avoid water.

129. hydrophobic interactions

Association of water-resisting molecules in an aqueous environment.

130. lecithin

A phospholipid characteristic of animal tissues.

131. lipid

Diverse group of organic substances soluble in organic solvents.

132. lipoprotein

A molecule made of lipid and protein.

133. monoglyceride

Monoacylglycerol.

134. nonpolar

Lacking charged regions.

135. phosphatidic acid

Component of a phosphoglyceride consisting of two fatty acids and a phosphate group.

136. phosphoglyceride

Main membrane component made of glycerol, two fatty acids, and phosphate.

137. phospholipid

A lipid made of glycerol, fatty acids, and phosphoric acid.

138. **polar compound**

A molecule having a charged area or polarity.

139. **saturated fatty acid**

A fatty acid lacking double bonds in the carbon chain and being saturated with hydrogen.

140. **saturation**

Condition of having all chemical affinities satisfied.

141. **sphingolipid**

Lipid containing sphingosine, an amine alcohol, capable of binding polar groups.

142. **sphingomyelin**

Sphingolipid containing phosphatidyl ethanolamine or phosphatidyl choline as polar group.

143. **sphingosine**

Long hydrocarbon chain with amine alcohol; backbone of sphingolipid.

144. **steroid**

A lipid with a complex four-ring structure.

145. **triglyceride**

A triacylglycerol (glycerol and three fatty acids).

146. **unsaturated fatty acid**

Fatty acid with pairs of hydrogen atoms replaced by double bonds in the carbon chain.

Proteins

147. **alpha helix**

Spiral secondary structure of a protein.

148. **amino acid**

Building block of a protein having at least one carboxyl and one amino group.

149. **amphipathic molecule**

A molecule with spatially separated hydrophobic and hydrophilic regions.

150. **beta pleated sheet**

Sheetlike secondary structure of a protein in which polypeptide chains are linked by hydrogen bonds.

151. **colinear**

Having similar parts arranged in the same order.

152. **collagen**

Strong fibrous protein found in animal connective tissues.

153. **deamination**

Removal of an amino group from a molecule.

154. **denaturation**

An alteration in the shape and properties of a protein molecule.

155. **essential amino acid**

Amino acid required in the diet because the body cannot make it.

156. **extensin**

Protein in cell walls of plants and fungi.

157. fibrous protein

Protein having a highly ordered alpha helix or beta pleated sheet structure.

158. globular protein

Protein with irregular secondary structure.

159. glycine

An amino acid with the simplest chemical structure.

160. glycoprotein

A molecule that contains both carbohydrate and protein components.

161. heme

Iron-containing porphyrin that can be oxidized or reduced while bound to a protein; the heme in hemoglobin.

162. hemoglobin

Protein containing heme that carries oxygen and carbon dioxide in blood.

163. integral membrane protein

Hydrophobic protein on interior surface of a membrane with hydrophilic parts protruding on one/both membrane sides.

164. iron-sulfur protein

Protein containing iron, sulfur, and cysteine groups that serves as an electron carrier.

165. lectin

Protein from plant seeds that binds sugars and is used to study glycoproteins.

166. monomeric protein

Protein consisting of a single polypeptide chain.

167. mucoprotein

Protein containing acidic carbohydrate elements found in the fuzzy layer.

168. multimeric protein

Protein consisting of two or more polypeptide chains.

169. oxidative deamination

Removal of an amino group with concurrent oxidation of molecule to a keto-acid.

170. peripheral membrane protein

Hydrophilic protein found on the surface of a membrane.

171. phospholipid transfer protein

Protein that mediates transfer of a specific phospholipid from one membrane to another.

172. polypeptide

A chain of amino acids held together by peptide bonds.

173. primary structure

Sequence of amino acids in a polypeptide or nucleotides in a nucleic acid.

174. protein

A polymer of amino acids.

175. quaternary structure

Protein structure involving interconnections between two or more polypeptides.

176. random coil

Region of irregular secondary structure of a protein.

177. renaturation

Restoration of normal shape of a denatured protein.

178. secondary structure

Protein structure produced by interactions along polypeptide chain; helix or pleated sheet.

179. specificity

The attribute of being specific.

180. terminal glycosylation

Change in glycoproteins in Golgi apparatus by addition or deletion of sugars to carbohydrate chains.

181. tertiary structure

Protein structure involving interactions between distant amino acids and three-dimensional folding.

182. transamination

Transfer of an amino group from one molecule to another.

183. turnover

Reuse of a substance made available by a catabolic reaction.

Enzymes

184.   **activation energy**

Energy needed to start a chemical reaction.

185.   **active site**

Location on an enzyme molecule at which substrate binds and reaction takes place.

186.   **allosteric effector**

Small molecule that changes shape of an allosteric protein by binding to a site other than the active site.

187.   **allosteric enzyme**

Enzyme that can change from active to inactive shape, depending on whether effector is bound to it.

188.   **allosteric regulation**

Turning on and off of a reaction pathway by controlling the shape of an allosteric enzyme.

189.   **amide bond**

Covalent bond between carboxyl group of one molecule and amino group of another.

190.   **aminopeptidase**

An exopeptidase that removes amino acids from the amino terminus of a peptide chain.

191.   **amylase**

An enzyme that digests starch.

192.   **autolysis**

Self-destruction by a cell's own hydrolytic lysosomal enzymes.

193.   **biosynthesis**

Creation of new molecules by chemical processes in living organisms.

194. **bond energy**

Energy required to break a particular chemical bond.

195. **calorie**

Quantity of heat needed to raise the temperature of one gram water one degree Celsius.

196. **carboxypeptidase**

An enzyme that acts outside cells to digest peptide chains from the carboxyl terminus.

197. **catalyst**

A substance that increases a chemical reaction rate.

198. **coenzyme**

A substance that works with an enzyme in activating chemical reactions.

199. **competitive inhibition**

Enzyme inhibition by inhibitor that competes with substrate for active site.

200. **condensation reaction**

Reaction in which two molecules are joined and water formed.

201. **dehydration**

Removal of water.

202. **effector site**

Region of an allosteric protein where effector binds and alters configuration and activity.

203. **endopeptidase**

Enzyme that degrades a small peptide chain by breaking internal peptide bonds.

204. **enzyme**

A protein that increases the rate of a chemical reaction in a living organism.

205. **enzyme kinetics**

Assessment of enzyme reaction rates and factors that affect them.

206. **equilibrium constant**

Ratio of product to reactant for a reaction at equilibrium.

207. **exopeptidase**

Enzyme that degrades small peptide chains from one end or the other.

208. **holoenzyme**

A complete enzyme.

209. **hydrolase**

Enzyme that splits molecules by hydrolysis.

210. **hydroxylation reaction**

Reaction in which oxygen is used to make an -OH group on an organic compound.

211. **induced-fit model**

Model of how enzymes become more fitted to their substrates after they bind to them.

212. **inducible enzyme**

Enzyme that is synthesized only in the presence of its substrate.

213. **inhibition**

Preventing from occurring.

214. **irreversible inhibitor**

Molecule that binds to and permanently inactivates an enzyme.

215. **kinase**

Enzyme that adds a phosphate group to a molecule.

216.   Lineweaver-Burk equation

Linear equation from inverting Michaelis-Menten equation; used to analyze enzyme inhibition.

217.   lock-and-key model

Model of enzyme function in which fit between enzyme and substrate is analogous to fit of lock and key.

218.   lysozyme

An enzyme in tears that can destroy microbes.

219.   maximum velocity

Fastest rate approached by enzyme-catalyzed reaction as the substrate concentration increases.

220.   Michaelis constant

Substrate concentration at which an enzyme-catalyzed reaction proceeds at one-half maximum velocity.

221.   Michaelis-Menten equation

Relationship between substrate concentration and reaction velocity.

222.   noncompetitive inhibition

Enzyme inhibition from reversible binding of inhibitor to a site other than the active site.

223.   peptidase

Enzyme that breaks down peptides by breaking peptide bonds.

224.   permease

Enzyme imbedded in membrane that mediates passage of substances across the membrane.

225.   phosphoester bond

Ester bond formed by removing -OH from a phosphate and -H from an alcohol.

226.   phosphorylation

Binding of a phosphate group to a molecule.

227. **prosthetic group**

Small metallic or organic molecule that contributes to catalytic activity of an enzyme.

228. **protease**

Enzyme that breaks peptide bonds and releases small peptides and individual amino acids.

229. **proteolysis**

Protein degradation by hydrolysis of peptide bonds.

230. **reversible inhibitor**

Inhibitor that inactivates an enzyme when bound to it by reversible binding.

231. **secondary electron**

One of many electrons used to form an image in scanning electron microscopy.

232. **substrate activation**

Property of active site of an enzyme that renders its substrate maximally reactive.

233. **substrate analog**

Molecule that closely resembles normal substrate and binds to substrate but cannot undergo reaction.

234. **synthetic work**

Energy use to make new molecules or new bonds.

235. **terminal oxidase**

Enzyme that can transfer electrons directly to oxygen.

236. **triple bond**

Bond between atoms in which three pairs of electrons are shared.

Organelles

237. amyloplast

A plastid that stores starch.

238. axoneme

A central shaft of nine outer doublets and a central pair of microtubules in a flagellum or cilium.

239. cell coat

Glycoproteins and other materials found exterior to the cell membrane in animal cells.

240. cell wall

Rigid, nonliving cellulose containing structure outside cell membrane in plants, bacteria, algae, and fungi.

241. central vacuole

In plant cells, large membrane-bound organelle that helps to maintain turgor and can store and/or degrade materials.

242. centriole

One of a pair of intracellular bodies that participate in forming a mitotic spindle.

243. chloroplast

A membrane-bound organelle containing chlorophyll and enzymes needed for photosynthesis.

244. chromoplast

Plastid with pigments that give color to plant parts.

245. cilium

A tiny hairlike projection found on some epithelial cells.

246. cisterna

Flat, membrane-bound sac, which is part of an organelle.

247.   cisternal space

Region inside a cisterna.

248.   condensing vacuole

Vacuole from Golgi apparatus that loses water to form a secretory granule.

249.   cortical cytoplasm

Cytoplasm lying beneath a cell membrane.

250.   crista

Infolding of inner mitochondrial membrane, which contains enzymes of oxidative phosphorylation.

251.   cytoskeleton

The organelles forming a cell's internal framework.

252.   cytosol

The fluid part of cytoplasm that suspends organelles.

253.   dictyosome

Stack of cisternae of Golgi apparatus.

254.   endoplasmic reticulum

A membranous vesicular network within a cell.

255.   flagellum

A movable hairlike process on a cell.

256.   forming face

Face of Golgi apparatus typically oriented toward the endoplasmic reticulum.

257.   gel

A liquid state in a colloidal dispersion.

258. **glyoxysome**

Membrane-bound organelle with enzymes that convert stored fat to carbohydrate in germinating seeds.

259. **Golgi apparatus**

Membranous vesicles clustered in cells that complete synthesis of secretions.

260. **Golgi-associated ER**

Part of ER that borders on Golgi apparatus and makes lysosomes.

261. **inner membrane**

Inner of two membranes surrounding the matrix of a mitochondrion.

262. **intermediate filament**

Filamentous protein that forms part of the structure of cytoskeleton of eukaryotic cells.

263. **intermembrane space**

Region of a mitochondrion or chloroplast between inner and outer membranes.

264. **keratin filament**

Protein filament in epithelial cells, which cover body surfaces and line cavities.

265. **large ribosomal unit**

Part of a ribosome with a hight sedimentation coefficient.

266. **leaf peroxisome**

Membrane-bound organelle containing enzymes of photorespiration.

267. **lipid body**

Organelle that stores fats in plant cells.

268. **lysosome**

Membrane-bound organelle that contains digestive enzymes.

269. matrix

Semifluid region inside a cell or organelle.

270. maturing face

Concave face on Golgi apparatus usually placed toward cell surface.

271. microbody

Peroxisome.

272. microfibril

Aggregate of long cellulose rods that strengthen cell walls of plants and fungi.

273. microfilament

A small, hollow protein fiber in cytoplasm that aids in movement or forms part of a cytoskeleton.

274. micromere

Small cell at end of vegetal hemisphere in sea urchin embryo.

275. microsome

Vesicle made of fragments of endoplasmic reticulum in cell homogenate.

276. microtrabecular lattice

Network of filaments that appear to interconnect elements of the cytoskeleton.

277. microtubule

A cylindrical organelle that forms part of a cell's mitotic spindle.

278. microtubule-associated protein (MAP)

Protein that binds along length of microtubule and enhances polymerization of tubulin subunits.

279. microvillus

Projections of membrane from surface of cells, especially those that absorb nutrients.

280. mitochondrion

An organelle that contains enzymes for oxidative and energy-capturing processes.

281. neurofilament

Fiber found in a neuron.

282. organelle

A tiny functional unit within a cell.

283. outer membrane

Outer membrane layer surrounding a mitochondrion or chloroplast.

284. palisade cell

Columnar leaf cell where photosynthesis occurs.

285. peptidoglycan

Bacterial cell wall component consisting of polysaccharide chains cross-linked with peptides.

286. peroxisome

An organelle containing oxidative enzymes.

287. plasmodesma

Channel through pores in cell walls of two adjacent plant cells that allows communication between cells.

288. plastid

Organelle of a plant cell that can become a chloroplast, chromoplast, or amyloplast.

289. primary lysosome

Digestive organelle with full set of hydrolytic enzymes not yet involved in digestion.

290. protofilament

Linear polymer of tubulin subunits that forms the structure of microtubules.

291.   ribosomal subunit

One of ribonucleoprotein particles that combine to form functional ribosome.

292.   ribosome

An organelle containing ribonucleic acid and protein where protein synthesis occurs.

293.   rough endoplasmic reticulum

ER with ribosomes bound to its surface and that participates in protein synthesis.

294.   second messenger

Intracellular mediator of extracellular messages.

295.   secondary lysosome

Organelle resulting from fusion of lysosome and phagocytic vacuole.

296.   secretory granule

Membrane-bound vacuole in eukaryotic cell that carries proteins from Golgi apparatus to plasma membrane.

297.   secretory protein

Protein that will be secreted from the cell that made it.

298.   sedimentation coefficient (S)

Rate at which a cell component or macromolecule moves in a centrifugal force field.

299.   self-assembly

Property of containing information required to establish structure of a particle.

300.   semiautonomous organelle

Organelle that contains DNA and can encode some of its own polypeptides.

301.   small ribosomal unit

Ribosome component with low sedimentation coefficient that combines with large subunit to form ribosome.

302. **smooth endoplasmic reticulum**

ER lacking ribosomes that participates in lipid synthesis and packaging of secretory proteins.

303. **sol**

A liquid state of a colloidal dispersion.

304. **stroma lamella**

One of several membranes that connect thylakoid disks of a chloroplast.

305. **structural macromolecule**

Polymer of one or several subunits that gives mechanical strength to a structure.

306. **supramolecular structure**

Cell component made of macromolecules in complex assembly.

307. **tau protein**

Microtubular protein that binds along tubule and enhances tubulin polymerization.

308. **tubulin**

A protein that forms intracellular microtubules.

309. **vacuole**

Membrane-bound organelle that temporarily stores or transports substance in a cell.

310. **zymogen granule**

Secretory granule containing precursors of digestive enzymes.

Nuclear Structures

311. **central granule**

Plug-like structure in annulus of a nuclear pore.

312. **chromatin**

Nuclear material that condenses into distinct chromosomes during cell division.

313. **chromatin fiber**

Element of eukaryotic chromosome with closely spaced nucleosomes.

314. **chromosome**

In a human cell, one of 46 nuclear structures made of DNA and protein.

315. **chromosome puff**

Segment of polytene chromosome undergoing transcription.

316. **inner nuclear membrane**

Inner of two unit membranes of the nuclear envelope in eukaryotic cells.

317. **kinetochore**

Chromosome region to which spindle microtubules attach.

318. **lampbrush chromosome**

Chromosome in egg nucleus with extended DNA loops where RNA is being synthesized.

319. **nuclear**

Of the nucleus.

320. **nuclear cortex**

Electron-dense fibrous material on nuclear side of inner nuclear membrane.

321. **nuclear envelope**

Double membrane around nucleus with many small pores.

322. **nuclear matrix filament**

Filament within a nucleus.

323. **nuclear organizer region (NOR)**

Chromosomal site where genes for rRNA are located and nucleoli form.

324. **nuclear pore**

Opening in the nuclear envelope large enough for small molecules to pass between nucleus and cytoplasm.

325. **nucleoid**

Cytoplasmic region where genetic material of prokaryote is located.

326. **nucleolus**

A body containing RNA within a nucleus.

327. **nucleoplasm**

The substance of a nucleus.

328. **outer nuclear membrane**

Outer of two unit membranes of nuclear envelope.

329. **sex chromosome**

A chromosome associated with maleness or femaleness; X or Y chromosome in mammals.

330. **spindle fiber**

Microtubules in eukaryotic cells involved in the movement of chromosomes during mitosis and meiosis.

Membranes

331. **activated monomer**

Monomer having a higher energy state for being linked to a carrier molecule.

332. **annulus**

Connecting cytoplasm in a plasmodesmata between two plant cells through which materials pass between the cells.

333. **autophagic lysosome**

A lysosome that digests substances from within the cell.

334. **autophagic vacuole**

A membrane-bound sac containing an organelle or other cell part in the process of being digested.

335. **bacteriorhodopsin**

Bacterial membrane protein that works with rhodopsin to transport protons across the membrane.

336. **belt desmosome**

Continuous tight adhesion between adjacent plasma membranes.

337. **binding site**

A site where a particular molecule binds to a membrane or other structure.

338. **cell junction**

Modified plasma membranes at site of contact between adjacent animal cells.

339. **cell membrane**

Lipid and protein compounds that form the boundary of a cell.

340. **clathrin**

Large protein that lines coated pits or covers coated vesicles and participates in intracellular transport.

341. coated pit

Plasma membrane region lined with clathrin that invaginates during endocytosis to form coated vesicle.

342. desmosome

Region of tight adhesion between adjacent animal cells.

343. desmsotubule

Passage in the central channel of a plasmodesma between two plant cells.

344. electrochemical gradient

Gradient of charged particles across a membrane.

345. endosome

Cytoplasmic vesicle of eukaryotic cell resulting from fusion of small vesicles from endocytosis.

346. fluid-mosaic model

A model of molecular arrangements in a cell membrane.

347. fluidity

Property of a membrane that allows proteins and lipids to move laterally.

348. food cup

Membrane that surrounds food particles taken in by phagocytosis in protozoa.

349. gap junction

Region of close proximity of two adjacent plasma membranes across which chemicals and electrical signals pass.

350. glycocalyx

Combined cell coat and fuzzy layer around an animal cell.

351. glycosaminoglycan

Carbohydrate with repeating disaccharide units found in fuzzy layer of animal cells.

352.   **heterophagic lysosome**

Lysosome containing hydrolytic enzymes that digest matter of extracellular origin.

353.   **ion carrier**

Substance that surrounds an ion with a hydrophobic coat so it can diffuse through hydrophobic interior of membrane.

354.   **ionophore**

Substance that increases permeability of a membrane for specific ions.

355.   **ligand**

That which binds to a receptor.

356.   **lipid bilayer**

Unit of membrane structure with two layers of phospholipids with heads toward aqueous environment.

357.   **lipid monolayer**

Single layer of lipid molecules with hydrophilic ends toward polar and hydrophobic ends in nonpolar environment.

358.   **membrane**

Selectively permeable barrier around a cell or organelle.

359.   **membrane asymmetry**

Membrane property related to differences in proteins associated with inner and outer surfaces.

360.   **membrane turnover**

Removal and replacement of lipids and proteins in a membrane.

361.   **perinuclear space**

Fluid-filled space between the two nuclear membranes.

362.   **perinucleolar chromatin**

Chromatin fibrils surrounding the nucleolus and containing DNA that codes for rRNA.

363. **phagaocytic vesicle**

Membrane-bound sac containing material taken in by a phagocyte.

364. **phagocyte**

Cell that carries out phagocytosis.

365. **plasma membrane**

Membrane forming the boundary of a cell.

366. **polymorphonuclear leukocyte**

White blood cell that has an irregular lobed nucleus.

367. **primary cell wall**

Wall of a growing plant cell.

368. **receptor**

A specific site with which a specific substance can bind; cell that responds to signals sensed from the environment.

369. **residual body**

Secondary lysosome containing undigestible material.

370. **secondary cell wall**

Wall formed on inner surface of primary wall after plant cell has reached full size and final shape.

371. **spot desmosome**

Tight adhesion point between plasma membranes of adjacent cells.

372. **tight junction**

Union of plasma membranes of adjacent animal cells extending around the circumference of the cell.

373. **tonofilament**

Filament that increases tensile strength of some cell junctions.

374.   transport vesicle

Coated vesicle from rough endoplasmic reticulum that trans-fers proteins and lipids to Golgi apparatus or other site.

375.   unit membrane

Structural entity consisting of a phospholipid bilayer with imbedded and attached proteins.

376.   vesicle

Spherical membrane-bound structure.

Membrane Functions

377. **activation**

Increasing a molecule's potential for reactivity.

378. **active transport**

Transport of a substance against a gradient using a carrier molecule, enzyme, and cellular energy.

379. **adsorptive endocytosis**

Entry of a substance into a cell after attaching to the cell membrane.

380. **autophagy**

Digestion within a cell of worn or unneeded organelles or other cell parts.

381. **buoyant density**

Point in gradient at which density of molecule or organelle equals density of surrounding solution.

382. **cellular secretion**

Process that makes and releases substances from cells.

383. **cellular transport**

Process that exchanges substances between cell and environment.

384. **coated vesicle**

Membrane-bound structure covered with clathrin involved in intracellular transport.

385. **connexon**

Doughnut-shaped aggregation of protein subunits that forms a channel through plasma membranes at gap junctions.

386. **coupled transport**

Concurrent facilitated transport of two substances across a membrane in the same or opposite directions.

387. density gradient

Presence of variation in solute concentrations in a solution.

388. diffusion

Random molecular movement that results in net solute movement from its higher to its lower region of concentration.

389. digitalis

Plant steroid that blocks sodium-potassium pump in animal cells.

390. drug detoxification

Modification of a drug to increase its rate of removal from the body.

391. endocytosis

Entry of material into a cell by membrane surrounding it and carrying the material into the cell as it infolds.

392. exocytosis

Fusion of vesicles with plasma membrane and expulsion of vesicle from cell.

393. extracellular

Outside a cell.

394. extracellular digestion

Breakdown of components outside a cell by enzymes released from the cell.

395. facilitated diffusion

Diffusion down a gradient on a carrier molecule but not requiring cellular energy.

396. facilitated transport

Movement of substances across a membrane with the aid of permeases imbedded in the membrane.

397. filtration

Passage of a fluid across a membrane by mechanical pressure.

398. **gradient**

The rate of change in the magnitude of concentration, pressure, or other variable.

399. **hydrostatic pressure**

Force exerted by a fluid.

400. **hyperosmotic**

Having higher osmotic pressure than a reference solution.

401. **hypertonic**

Causing movement of water out of cells.

402. **hyposmotic**

Having lower osmotic pressure than a reference solution.

403. **hypotonic**

Causing movement of water into cells.

404. **intracellular**

Within a cell.

405. **intracellular transport**

Movement of substances across membranes of organelles inside a cell.

406. **isosmotic**

Having the same osmotic pressure as a reference solution.

407. **isotonic**

Causing no net water movement across a cell membrane.

408. **lateral diffusion**

Movement of molecules within the plane of a membrane.

409. macropinocytosis

Mechanism by which plasma membrane invaginates around and ingests a vesicle 1 to 2 microns in diameter.

410. micropinocytosis

Uptake of vesicles with diameter no greater than 1 micron by invagination of plasma membrane.

411. osmolarity

A solution's osmotic concentration determined by the number of osmotically active particles it contains.

412. osmosis

Diffusion of water through a membrane from its own higher to a lower concentration.

413. osmotic pressure

Pressure created by osmosis.

414. oubaine

Toxic steroid from plants that inhibits sodium-potassium pump in animal cells.

415. passive transport

A process that moves substances without energy expenditure by the organism.

416. phagocytosis

Endocytosis in which hareg particles (even whole cells) are engulfed and digested.

417. phosphorylating transport

Transfer of a sugar into a bacterium with phosphorylation a part of the uptake mechanism.

418. pinocytosis

Kind of endocytosis in which dissolved substances are taken into vacuoles; cell drinking.

419. plasmolysis

The process by which a plant cell protoplast shrinks from its cell wall due to water loss in a hypertonic solution.

420. **proton motive force**

Electrochemical force due to difference in proton concentrations across membrane.

421. **receptor-mediated endocytosis (RME)**

Endocytosis initiated at coated pits that produces coated vesicles; adsorptive pinocytosis.

422. **selective permeability**

A property of membranes allowing passage of some substances while preventing the passage of others.

423. **simple active transport**

Transport of a solute across a membrane in one direction without concurrent flow of other substances.

424. **sodium cotransport**

Exergonic inward sodium transport used to drive active transport of organic solutes.

425. **sodium-potassium pump**

Mechanism that actively moves Na ions out of cells and K ions into them against gradients.

426. **surface tension**

Resistance to rupture by the surface film of a liquid.

427. **surface-to-volume ratio**

The surface area of a structure divided by its volume.

428. **tonicity**

The degree to which fluid can move into or out of cells.

429. **transcellular transport**

Movement of substances across a cell, usually accompanied by entrance on one side and exit on the other.

430. **transverse diffusion**

Movement of phospholipid or protein from one layer of a membrane to the other.

431. vectorial pumping

Movement of molecules in one direction across a membrane.

Energy

432.   **adenosine triphosphate**

An important energy storage molecule.

433.   **amphibolic pathway**

A sequence of reactions that can serve both to break down substances and provide others for synthetic pathways.

434.   **anabolic**

Of anabolism.

435.   **anabolic pathway**

A sequence of reactions that produces more than one kind of cellular component.

436.   **anabolism**

Synthetic, energy using process.

437.   **bioenergetics**

Study of how thermodynamics apply to processes in living organisms.

438.   **bioluminescence**

Light production by reacting ATP with luminescent substance.

439.   **catabolic pathway**

Reactions that degrade cellular substances.

440.   **catabolism**

Breakdown of molecules that makes energy available.

441.   **closed system**

System that exchanges energy but not matter with surroundings.

442. **concentration work**

Energy use to move molecules or ions across a membrane against a gradient.

443. **covalent bond**

A chemical bond formed by shared electrons between two atoms.

444. **disulfide bond**

Covalent bond between two sulfur atoms.

445. **double bond**

Chemical bond consisting of two pairs of shared electrons.

446. **electrical work**

Energy used to transport charged particles against a concentration gradient.

447. **endergonic**

Requiring energy, as in a chemical reaction.

448. **energy**

Capacity to do work.

449. **entropy**

Tendency toward chaos or disorder.

450. **ester bond formation**

Condensation by which -OH of a carboxyl group and -H of an alcohol group are removed and a bond formed.

451. **exergonic**

Releasing energy, as in a chemical reaction.

452. **first law of thermodynamics**

Energy can be converted from one form to another but neither created nor destroyed.

453. free energy

Thermodynamically, energy that can be extracted from a molecule and used to do work.

454. free energy change

Free energy liberated or required by a reaction or process.

455. glycosidic bond

Bond that links a sugar to another sugar or other molecule.

456. heat

Transfer of energy as temperature changes.

457. heat of vaporization

Amount of energy required to convert a substance from liquid to gaseous phase.

458. hydrogen bond

Weak covalent bond between hydrogen and another element, such as oxygen or nitrogen.

459. ionic bond

A chemical bond with atoms held together by the attraction of unlike charges.

460. isothermal

Having a constant temperature.

461. kilocalorie

Energy needed to raise the temperature of 1 kg of water 1 degree C.

462. mechanical work

Energy use to produce a physical change in location of cell or its parts.

463. metabolic pathway

Sequence of enzymatic reactions that convert one molecule to another by way of several intermediates.

464. **metabolism**

All chemical reactions in a living organism.

465. **metabolite**

Substance that participates in a metabolic pathway.

466. **negative regulation**

Regulation of a metabolic pathway by an end product slowing or stopping reactions.

467. **open system**

System that exchanges both energy and matter with surroundings.

468. **peptide bond**

A chemical bond between the amino group of one amino acid and the carboxyl group of another.

469. **positive regulation**

Regulation of a metabolic pathway by the presence of a substance such as a substrate.

470. **second law of thermodynamics**

Physical and chemical changes proceed in a manner that increases entropy in the universe.

471. **specific heat**

The amount of heat needed to increase the temperature of a specific volume of substance one degree Celsius.

472. **spontaneous**

Property of chemical reactions that release energy and therefore can occur without energy input.

473. **steady state**

Nonequilibrium condition in open system with matter flowing and all components present in changing amounts.

474. **surroundings**

Part of universe not included in a system under study.

475.   thermodynamics

Study of laws governing energy changes in chemical or physical processes.

476.   uridine triphosphate (UTP)

A high energy molecule.

477.   work

Energy transfer from one place or form to another by any process other than heat flow.

Energy-Capturing Metabolism

478. acetyl coenzyme A

Ester of acetic acid and coenzyme A that enters Krebs cycle.

479. adenosine diphosphate (ADP)

Adenosine with two phosphate groups attached.

480. adenosine monophosphate (AMP)

Adenosine with one phosphate group attached.

481. aerobic

In the presence of oxygen.

482. alcoholic fermentation

Anaerobic breakdown of carbohydrate to form ethanol and carbon dioxide.

483. anaerobic

Lacking oxygen.

484. ATP synthetase complex

Complex of particles that leads to ATP synthesis by linking the flow of protons to ATP generation in photosynthesis.

485. beta oxidation

Reaction sequence by which fatty acids are broken down to acetyl CoA.

486. chemisomotic model

Electrochemical gradient across inner mitochondrial membrane links electron transport and ATP generation.

487. citric acid cycle

A sequence of reactions that oxidize acetyl-CoA; Krebs cycle; tricarboxylic acid cycle.

488. **coenzyme A (CoA)**

Molecule that forms high-energy bond with and transports acyl group.

489. **coenzyme Q (CoQ)**

Component of electron transport chain that receives electrons from NAD and FAD.

490. **Cori cycle**

Movement of lactate from muscle to liver and glucose from liver to muscle.

491. **cyclic AMP (cAMP)**

AMP arranged in ring structure; acts as cellular second messenger.

492. **cytochrome**

One of several heme-containing proteins that transfer electrons from CoQ to oxygen.

493. **cytochrome C oxidase**

Enzyme of electron transport system that transfers electrons to molecular oxygen.

494. **dehydrogenation**

Removal of hydrogen, oxidation.

495. **electron transport**

Oxidation under aerobic conditions of electron carriers.

496. **electron transport system**

Enzymes and coenzymes in cristae of mitochondria that move electrons from substrates to oxygen.

497. **F-one particle**

ATP synthetase component that protrudes from an inner mito-chondrial membrane and acts as a site for ATP generation.

498. **F-zero particle**

ATP synthetase component that moves protons in the inner mitochondrial membrane.

499.   fermentation

An anaerobic metabolic process in which carbohydrate is broken down to alcohol and other simple molecules.

500.   flavin adenine dinucleotide (FAD)

A coenzyme that carries hydrogen.

501.   flavin mononucleotide (FMN)

Single nucleotide electron carrier in energy metabolism.

502.   flavoprotein

Molecule consisting of protein bound to derivative of riboflavin.

503.   gluconeogenesis

Metabolic pathway that makes glucose from noncarbohydrate substances.

504.   glycogenesis

Metabolic pathway for glycogen synthesis.

505.   glycogenolysis

Metabolic pathway for glycogen breakdown.

506.   glycolysis

Metabolic pathway for breakdown of glucose to pyruvic acid.

507.   glyoxylate cycle

Set of reactions that generate succinate from acetyl CoA required for synthesis of sugars from triacylglycerols.

508.   hydrogenation

Addition of hydrogen ions (protons) to molecules; reduction.

509.   lactate fermentation

Anaerobic breakdown of carbohydrates to lactate.

510. **lethal synthesis**

Making of a toxic substance from a nontoxic one.

511. **macronutrient**

Nutrient needed in relatively large amounts.

512. **micronutrient**

A nutrient needed in relatively small quantities.

513. **mineral**

Inorganic substance.

514. **NADH dehydrogenase**

Enzyme that transfers electrons from NADH to coenzyme Q.

515. **nicotinamide adenine dinucleotide (NAD)**

A coenzyme that transports hydrogen atoms or electrons in oxidation-reduction reactions.

516. **obligate aerobe**

Organism that must have free oxygen as an electron acceptor.

517. **obligate anaerobe**

Organism that must have something other than oxygen as an electron acceptor.

518. **oxidation-reduction couple**

Pair of molecules of which one is oxidized and the other is reduced in a chemical reaction.

519. **oxidative phosphorylation**

Capture of energy in ATP during oxidative metabolism.

520. **phosphodiester bond**

Bond that links two alcohols to a single phosphate group.

521.   phosphogluconate pathway

Reactions that rearrange pentose phosphates into hexose phosphates that can enter glycolysis.

522.   photorespiration

Use of light energy for oxidative metabolism of glycolate.

523.   respiration

Oxidation of organic molecules for energy.

524.   respiratory assembly

Functional unit of an inner mitochondrial membrane, containing electron transport molecules.

525.   respiratory control

Regulation of oxidative phosphorylation and electron transport by the quantity of ADP available.

526.   substrate-level phosphorylation

Use of ADP and inorganic phosphate to form ATP as a part of a reaction in a metabolic pathway, not in mitochondrion.

527.   system

Portion of universe being considered.

528.   tautomerization

Movement of a proton that changes chemical properties of a molecule.

529.   uncoupler

Substance that separates link between electron transport and phosphorylation.

Photosynthesis

530. **absorption spectrum**

Relative degree to which a pigment absorbs different wavelengths of light.

531. **accessory pigment**

Molecule with light-gathering properties that augments chlorophyll and may give color to plant tissue.

532. **action spectrum**

Relative degree to which different wavelengths of light affect a light-dependent process.

533. **bacteriochlorophyll**

A kind of chlorophyll found in bacteria that can receive electrons from sources other than water.

534. **bundle sheath cell**

Internal leaf cell near vascular bundle where carbon metabolism by way of the Calvin cycle occurs.

535. **C-four plant**

Plant in which carbon is fixed in 4-carbon compound.

536. **C-three plant**

Plant in which carbon is fixed in 3-carbon compound.

537. **Calvin cycle**

Series of reactions that fix carbon from carbon dioxide and subsequently synthesize sugars.

538. **carotenoid**

Yellow-orange plant pigment that absorbs violet to green light.

539. **CF-one**

Part of ATP synthetase in chloroplast that generates ATP.

540. CF-zero

Part of ATP synthetase in thylakoid membrane that moves protons.

541. chlorophyll

A green pigment capable of capturing light energy.

542. cyclic electron flow

Transfer of electrons from photosystem I with energy made available for ATP synthesis.

543. cyclic photophosphorylation

The capture of energy in chloroplasts with the formation of ATP through the activity of the cell's cytochrome system.

544. dark reaction

A part of photosynthesis that can occur in either light or dark and that transfers energy from light reactions.

545. Emerson enhancement effect

Greater photosynthetic activity produced with red light of two slightly different wavelengths.

546. glycolate pathway

Photosynthetic reactions by which phosphoglycolate is made in chloroplasts.

547. granum

A stack of thylakoids within a chloroplast.

548. intrathylakoid space

Space within membranes of thylakoids and lamellae of stroma.

549. light reaction

Events in photosynthesis that capture energy and occur in light.

550. mesophyll cell

Outer cell in C-four plant leaf; site of carbon fixation.

551. **nicotinamide adenine dinucleotide phosphate**

Coenzyme that transfers electrons in the Calvin cycle and other metabolic pathways.

552. **noncyclic electron flow**

Electron flow from water to NADP using light for energy.

553. **noncyclic photophosphorylation**

The capture of energy in chloroplasts with the formation of ATP and reduced NADP.

554. **oxygenic photoautotroph**

Organism in which water is the electron donor in photosynthesis.

555. **P/O ratio**

Ratio of ATP molecules generated per oxygen atom reduced.

556. **photochemical reduction**

Transfer of light-excited electrons from one molecule to another.

557. **photoexcitation**

Excitation of an electron by absorption of light energy.

558. **photolysis**

Use of light energy to oxidatively split water molecules.

559. **photophosphorylation**

The addition of phosphate groups and high energy bonds to a molecule during the capture of light energy.

560. **photoreduction**

Use of light energy to generate NADPH by transfer of excited electrons from chlorophyll to NADP via electron carriers.

561. **photosynthesis**

The process by which organisms capture light energy from the environment and store it in a usable form.

562. **photosynthetic unit**

Group of up to 300 chlorophyll molecules of which only a few molecules participate in photochemical reactions.

563. **photosystem**

Functional system of chlorophyll and other pigments embedded in thylakoid membrane.

564. **photosystem I**

System in which chlorophyll absorbs 700nm red light maximally after which electrons reduce NADP to NADPH.

565. **photosystem II**

System in which chlorophyll absorbs 680nm red light maximally after which electrons donated by water are excited.

566. **phycobilin**

Pigment in red and blue-green algae that absorbs green-to-orange light and gives algae their color.

567. **pigment**

Substance that absorbs some light and reflects light of wavelengths corresponding to the color perceived.

568. **plastocyanin**

Protein containing copper that donates electrons to chlorophyll P700 of photosystem I in light reactions.

569. **plastoquinone**

Molecule with quinone component that participates in transfer of electrons between photosystems I and II.

570. **quantum requirement**

Amount of light energy in photons needed for specific change.

571. **reaction center**

Part of photosynthetic unit having chlorophyll molecules that intiate electron transfer.

572. **stoma**

A pore in lower leaf epidermis through which gases diffuse into and out of mesophyll spaces.

573.   thylakoid

Parallel flattened sacs that form part of the membrane structure of a chloroplast.

Nucleic Acids

574. **actinomycin D**

An inhibitor of RNA synthesis.

575. **adenine**

Nitrogenous base classified as a purine found in nucleic acids.

576. **adenosine**

Compound consisting of adenine and ribose.

577. **B-DNA**

Right-handed DNA helix.

578. **base pairing**

Bonding of complementary purines and pyrimidines in double-stranded nucleic acids.

579. **Chargaff's rules**

In double stranded DNA, the number of adenines and thymines and the number of cytosines and guanines are equal.

580. **cohesive end**

Single-stranded DNA fragment from cleavage by a restriction enzyme that can attach to another similar fragment.

581. **consensus sequence**

Nucleotide order in highly conserved segment of DNA.

582. **constitutive heterochromatin**

Permanently condensed genetically inactive chromosomal region.

583. **core particle**

Histone octamer having DNA wound around it.

584. cytosine

Pyrimidine present in nucleotides.

585. deoxyribonuclease

An enzyme that digests DNA.

586. deoxyribonucleic acid (DNA)

A nucleic acid in chromosomes that directs protein synthesis and transmits genetic information to a new generation.

587. deoxyribose

Five-carbon monosaccharide in DNA.

588. depurination

Removal of a purine.

589. DNA glycosidase

Enzyme that finds and removes deaminated bases from DNA.

590. double helix

Double-stranded, spiral molecule as of DNA.

591. euchromatin

Diffuse, uncondensed, active chromatin.

592. exon

Group of nucleotides at the beginning of a molecule of RNA that persists into mature functional RNA.

593. facultative heterochromatin

Region of chromosome that has been specifically inactivated in a particular kind of cell.

594. guanine

A purine found in nucleotides of nucleic acids.

595. helix

Spiral shape of DNA and some other biopolymers.

596. histone

A kind of basic protein found in eukaryotic chromosomes.

597. insertional mutagenesis

Changing of transcriptional activity of nearby host genes by insertion of viral DNA.

598. major groove

Larger of grooves in double-helix DNA.

599. messenger RNA

A nucleic acid that carries information in the form of codons for the synthesis of a protein.

600. minor groove

Smaller of grooves between DNA strands in double helix.

601. negative supercoil

Coil in circular DNA by left-handed twist of relaxed molecule.

602. nonhistone chromosomal protein

Any of several acidic proteins found in small amounts in eukaryotic chromosomes.

603. nonreiterated sequence

DNA sequence found only once per haploid genome.

604. nucleic acid

A polymer of nucleotides; DNA or RNA.

605. nucleoside

Molecule containing a purine or pyrimidine and pentose sugar.

606. nucleosome

Structural unit of a chromosome containing 200 DNA base pairs associated with an octamer of histone proteins.

607. nucleotide

A molecule having a nitrogenous base, a 5-carbon sugar, and one or more phosphates.

608. packing ratio

Ratio of the length of a DNA molecule to the length of a chromosome into which it is packed.

609. palindromic

Reading the same forward and backward.

610. positive supercoil

Coil in circular DNA formed by right-handed twist of a relaxed molecule.

611. purine

A nitrogenous base with two rings found in nucleic acids.

612. pyrimidine

A nitrogenous base with one ring found in nucleic acids.

613. reiterated sequence

DNA sequence found in multiple copies in a haploid genome.

614. relaxed state

Circular DNA that lacks supercoils.

615. repair endonuclease

Enzyme that detects missing bases in DNA and breaks bonds at such sites so other repair enzymes can act.

616. repair synthesis

Removal of defective DNA segments and replacement with normal ones.

617. **replicon**

Self-replication DNA unit that includes site at which replication began.

618. **restriction enzyme**

An endonuclease that cuts double stranded DNA at sites having specific nucleotide sequences.

619. **restriction site**

The site at which an endonuclease acts.

620. **ribonuclease**

Enzyme that breaks phosphodiester bonds in RNA.

621. **ribonucleic acid (RNA)**

A nucleic acid made from information in DNA that is involved in protein synthesis.

622. **ribosomal RNA (rRNA)**

A nucleic acid that forms part of a ribosome.

623. **spacer sequence**

Sequence of RNA nucleotides excised during processing.

624. **structural gene**

Sequence of nucleotides that code for a particular polypeptide.

625. **supercoil**

Twist in circular DNA that makes helix coil on itself.

626. **thymine**

Pyrimidine found in DNA that pairs with adenine.

627. **topoisomerase**

Enzyme that converts relaxed DNA to supercoiled DNA.

628.   uracil

A pyrimidine in RNA, which is coded by adenine in DNA.

DNA Replication

629. **bidirectional replication**

DNA replication that proceeds from two separate replication forks in two separate directions simultaneously.

630. **DNA gyrase**

Enzyme that unwinds DNA strands during replication.

631. **DNA ligase**

Enzyme that joins DNA fragments by phosphoester bonds.

632. **DNA polymerase**

Enzyme that adds succesive nucleotides to DNA strand.

633. **DNA replication**

Replication of DNA in which each strand serves as a template for the other.

634. **endoreplication**

Repeated replication of chromosomes with no intervening separation of daughter chromosomes.

635. **gyrase**

Enzyme that helps unwind DNA helix for replication.

636. **helicase**

Enzyme that uses energy from ATP to unwind DNA helix and expose it to replication enzymes.

637. **heterogeneous nuclear RNA**

Transcription product in eukaryotic nucleus that probably gives rise to messenger RNAs.

638. **lagging strand**

DNA strand that grows in 3' to 5' direction in discontinuous segments.

639. **leading strand**

The DNA strand that is replicated continuously.

640. **noncoding strand**

DNA strand that is not used as a template for RNA synthesis.

641. **Okazaki fragment**

Short nucleotide sequence made from lagging strand as DNA replication intermediate.

642. **origin of replication**

Site on chromosome at which DNA replication begins.

643. **primase**

RNA polymerase that makes RNA primers that initiate DNA replication.

644. **primer**

Short RNA sequence at 5' end of Okazaki fragment to which nucleotides are added in DNA synthesis.

645. **primer recognition factor**

Protein that causes synthesis of short RNA segment on lagging strand of DNA.

646. **primosome**

Protein complex associated with primase that contains recognition factors for primer synthesis.

647. **recombinant DNA**

DNA segments combined from two different organisms.

648. **replication**

Duplication.

649. **replication bubble**

Structure in replicating DNA at which two replication forks move away from a common site of origin of replication.

650. **replication fork**

Y-shaped site at which DNA replication is occurring.

651. **RNA polymerase**

Enzyme that adds nucleotides to RNA strand according to DNA template.

652. **semiconservative replication**

The replication of DNA in which each molecule consists of one new and one old strand.

Cell Division

653. **acrocentric chromosome**

Chromosome having its centromere near one end.

654. **anaphase**

A mitotic stage during which chromosomes move apart.

655. **anaphase I**

The stage of the first meiotic division in which chromosomes move to opposite poles of dividing cell.

656. **anastral mitosis**

A kind of mitosis seen in higher plants that lacks an astral arrangement of microtubules at the spindle poles.

657. **antimitotic drug**

Substance that blocks mitosis by interfering with one or more processes in cell division.

658. **aster**

Short microtubules at the ends of a spindle in a dividing animal cell.

659. **astral mitosis**

A kind of mitosis in higher animals that has an astral arrangement of microtubules at the spindle poles.

660. **cell cycle**

A repetitive sequence of events involving DNA replication and cell division.

661. **cell plate**

Precursor to cell wall and separation of daughter cells.

662. **centromere**

Site on a chromosome at which sister chromatids are attached prior to anaphase; also location of kinetochore.

663. chiasma

In prophase I, X-shaped profile of homologous chromatids.

664. chromatid

Replicated chromosome still joined to sister at centromere.

665. chromosomal fiber

Microtubules that attach chromosome by its kinetochore to the spindle.

666. cleavage

Mitotic divisions without increase in size in early animal embryos.

667. cleavage furrow

Groove along which daughter cells separate during cleavage.

668. colchicine

Plant alkaloid that prevents tubulin polymerization and thus microtubule formation.

669. contact inhibition

Property of normal cells by which contact between them arrests cell division; lacking in malignant cells.

670. contractile ring

Belt of actin microfilaments under plasma membrane that constricts cleavage furrow in animal cell division.

671. cytokinesis

Division of the cytoplasm that follows division of a nucleus.

672. daughter cell

A cell that results from cell division.

673. diakineses

Last stage in prophase I of meiosis in which chromosomes are maximally condensed and nuclear envelope is dispersed.

674. diplotene

Fourth stage of prophase I in meiosis in which bivalent chromosomes begin to pull away from each other.

675. division phase

Phase in cell cycle in which mitosis and cytokinesis occur.

676. G-1 phase

Phase in eukaryotic cell cycle after division and before DNA synthesis begins.

677. G-2 phase

Phase in eukaryotic cell between chromosome replication and cell division.

678. gamete

Haploid reproductive cell.

679. gametic meiosis

A kind of meiosis that produces haploid gametes.

680. gametogenesis

Production of gametes.

681. gametophyte

Haploid generation in life cycle that produces gametes.

682. generation time

Length of the cell cycle.

683. growth phase

Phase in which eukaryotic cell doubles in mass and duplicates its contents in preparation for division.

684. interphase

A cell cycle stage during which the cell is not dividing.

685.   karyotype

Arrangement of chromosomes from a cell in pairs and in a fixed order.

686.   leptotene

First stage of prophase I of meiosis during which chromosomes initially condense and coil.

687.   meiosis

Two successive nuclear divisions with one chromosome replication, which produces four haploid nuclei.

688.   metacentric chromosome

Chromosome with centromere near its center.

689.   metaphase

A mitotic stage during which chromosomes align along the equator of a cell.

690.   metaphase I

Second phase of first meiotic division in which bivalent chromosomes align on metaphase plate.

691.   mitosis

Nuclear division that produces two identical nuclei.

692.   mitotic index

Proportion of cells in a culture that are in mitosis at any given time.

693.   mitotic spindle

Arrangement of microtubules that separate chromosomes during mitosis.

694.   monolayer of cells

Layer of nonoverlapping cells as is usually formed by normal cells in culture.

695.   pachytene

Third stage of first prophase in meiosis, when synapsis has occurred.

696. phragmoplast

Cylindrical central structure in a dividing plant cell in which microtubules help to form cell plate.

697. polar fiber

Microtubule bundle that extends from spindle pole to equator of spindle.

698. prophase

First phase of mitosis in which chromosomes condense and become visible.

699. prophase I

First phase of first meiotic division over which chromosomes condense, become visible, and synapse.

700. restriction point

Point in G-1 phase of eukaryotic cell cycle after which division must proceed.

701. S phase

In eukaryotic cell cycle, phase during which DNA is synthesized.

702. sister chromatid

One of a pair of chromatids that detach in anaphase of mitosis.

703. spindle pole

Either end of mitotic spindle, which contains a centriole and in animal cells microtubules.

704. spore

Haploid product of meiosis in organisms with alternation of generations.

705. sporic meiosis

Meiosis that produces haploid cells called spores, which in turn develop into haploid organisms.

706. sporophyte

Diploid component of life cycle of organism that has alternation of generations.

707. **synaptonemal complex**

Structure that allows close pairing of homologous chromosomes and crossing over during prophase I.

708. **teleocentric chromosome**

Chromosome having centromere near one end.

709. **telophase**

The last mitotic stage during which nuclei reform.

710. **telophase I**

Final stage in first meiotic division during which chromosomes complete movement to poles of cell.

711. **tetrad**

Four attached chromatids formed by synapsis.

712. **trigger protein**

Protein that accumulates in G-1 phase of cell cycle and makes the cell enter the S phase and eventually divide.

713. **zygotene**

Second stage of first prophase of meiosis I in which homologous chromosomes begin to pair and synapse.

Genetics / Mutations

714. allele

One of two or more forms of a particular gene.

715. Ames test

Screening test for carcinogens based on ability of such to cause Salmonella to mutate and grow on medium provided.

716. backcross

In genetics, crossing of a heterozygote with a homozygous member of parental generation.

717. benign tumor

A self-limiting tumor that does not spread.

718. bivalent

Synapsed pair of homologous chromosomes.

719. cancer

Uncontrolled cell division usually with invasion of other tissues.

720. carcinogen

Agent that can cause cancer.

721. carcinoma

Cancer derived from epithelial cells.

722. chromosome mapping

Specifying the location and order of genes on a chromosome by studying crossover frequencies.

723. chromosome theory of heredity

Hereditary factors are located on chromosomes in a cell's nucleus.

724.   **clone**

Group of identical cells derived from a single parent cell.

725.   **crossing-over**

The exchange of corresponding segments of DNA during meiosis.

726.   **diploid**

Having two sets of chromosomes.

727.   **direct mutagen**

Agent that directly alters nucleotide sequence in DNA.

728.   **dominant**

Allele that is expressed whenever present.

729.   **F-one generation**

First generation of offspring from a cross of two parental strains.

730.   **F-two generation**

Second generation of offspring from a cross of two parental strains.

731.   **frameshift mutation**

Insertion or deletion of nucleotides in DNA such that reading frame of codons is garbled.

732.   **gene**

Functional unit of heredity; a site on a chromosome that transmits a particular hereditary characteristic.

733.   **gene amplification**

Selective replication of DNA sequences within a genome.

734.   **genetic code**

The three-base sequences in messenger RNA derived from a DNA template that determine amino acid order in proteins.

735. **genome**

An organism's whole complement of DNA.

736. **genomic control**

Regulation in eukaryotes that involves loss or amplification of specific information of the genome.

737. **genotype**

Genetic composition of an organism.

738. **haploid**

Having one of a pair of chromosomes.

739. **heterochromatin**

Regions of chromosomes that remain condensed during interphase.

740. **heterozygous**

Having different alleles for a given trait.

741. **homozygous**

Having like alleles for a given trait.

742. **hybrid**

Progeny of a cross of genetically different parents.

743. **indirect mutagen**

Agent that can be modified in a cell to become a mutagen.

744. **law of independent assortment**

Mendel's law that pairs of alleles of different traits segregate independently.

745. **law of segregation**

Mendel's law that two alleles of a trait segregate without influencing each other.

746.   linkage group

Set of genes in close proximity on chromosome that are inherited and assorted together.

747.   malignancy

A tendency to become more virulent; a cancerous growth.

748.   map unit

Unit of measure between gene loci based on crossover frequencies.

749.   Mendelian genetics

Understanding of inheritance based on Mendel's work.

750.   metastasis

The transfer of disease from one organ to another.

751.   mitogen

Substance that stimulates mitosis.

752.   mutagen

Substance or agent that can induce mutations.

753.   mutation

A heritable change in DNA that alters structure or function of offspring.

754.   nonsense mutation

Change in DNA that produces a stop codon where an amino acid codon previously existed.

755.   oncogene

A gene that contributes to the development of cancer.

756.   phenotype

Observable characteristics determined by genetic factors.

757. **polymorphy**

Having numerous alleles for each locus.

758. **proto-oncogene**

Cancer-causing gene in animal cell that can be picked up by and inserted into a retrovirus genome.

759. **pyrimidine dimer**

Bonding between adjacent pyrimidines in DNA that blocks replication and transcription.

760. **recessive**

Allele that is expressed only in absence of a dominant one.

761. **recombinant DNA technology**

Procedures used to combine DNA from two or more sources.

762. **recombination**

Production of new genotypes from exchange of genes between homologous chromosomes during synapsis.

763. **remission**

Abatement of disease symptoms or the period during which the abatement occurs.

764. **sarcoma**

Tumor derived from connective tissue or other tissue of mesodermal origin.

765. **site-specific mutagenesis**

Modification of protein product by changing one or more specific nucleotides in a gene.

766. **teratogen**

An agent that causes defective embryonic development.

767. **true-breeding**

Progeny that are homozygous and phenotypically identical to parents.

768.   tumor necrosis factor

A substance that causes degeneration and death of tumor cells.

769.   wild type

Normal, nonmutant form of an organism found in nature.

Protein Synthesis

770. **adaptive enzyme synthesis**

Regulation of concentration of an enzyme by regulating its rate of synthesis according to cellular needs.

771. **aminoacyl site**

Location on a ribosome at which tRNA carrying an amino acid binds.

772. **aminoacyl tRNA**

Transfer RNA with an amino acid bound to it by an ester bond.

773. **aminoacyl tRNA synthetase**

Enzyme that forms ester bond between tRNA and the amino acid it carries.

774. **anticodon**

A three-base sequence of transfer RNA that fits with a particular codon on messenger RNA.

775. **attenuation**

A process of regulating bacterial operons that participate in amino acid synthesis.

776. **attenuation site**

A sequence of nucleotides between promotor and structural genes in some operons, which lacks RNA polymerase action.

777. **catabolic activator protein (CAP)**

Regulatory protein that binds to cAMP and then to site near a promotor and enhances transcription in some operons.

778. **catabolite repression**

Action of an alternative substrate in decreasing production of enzymes for another catabolic pathway.

779. **central dogma**

Genetic information flows from DNA to RNA to proteins.

780.   cistron

Sequence of DNA nucleotides that carries information for a single polypeptide.

781.   cloning

Production of clones.

782.   coding strand

Strand of DNA that serves as template for RNA synthesis.

783.   codon

A three-base sequence in messenger RNA derived from DNA and specifying amino acid placement in a protein.

784.   complementary base pairing

Bonding between certain bases in nucleic acid strands.

785.   core glycosylation

Initial binding of carbohydrates to polypeptides of the endoplasmic reticulum.

786.   cotranslational transport

Movement of protein across a membrane as it is synthesized.

787.   degenerate

Attribute of genetic code in which a given amino acid has more than one codon.

788.   docking protein

Protein in membrane of rough ER that binds a ribosomal complex in some instances of blocked translation.

789.   effector

Organic molecule that binds to and regulates an allosteric enzyme or other molecule.

790.   end-product repression

Repression of synthesis of enzymes for a pathway in proportion to the quantity of product accumulated.

791. **feedback inhibition**

Inhibition of a first enzyme in a pathway by accumulation of end product.

792. **fibrillar component**

Ribosomal RNA transcript associated with nucleolar protein.

793. **glycosylation**

Addition of sugar to a polypeptide chain usually done in the endoplasmic reticulum.

794. **granular component**

Subunits of ribosomes undergoing maturation in a nucleolus.

795. **initiation codon**

Codon that places the first amino acid in a protein.

796. **intron**

Nucleotide sequence in RNA that is not functional; intervening sequence.

797. **leader peptide**

Peptide made by leader sequence of mRNA that codes for enzymes of a biosynthetic pathway.

798. **leader sequence**

Nucleotides at 5' end of mRNA that regulates transcription of structural genes of an operon.

799. **monocistronic mRNA**

Molecule of mRNA that directs the synthesis of one polypeptide.

800. **N-formyl methionine**

Methionine with formyl group attached found as first amino acid in every polypeptide made by a prokaryote.

801. **nascent polypeptide**

Recently synthesized polypeptide still bound to ribosomes.

802. negative control

Regulation by turning off expression of an operon.

803. peptidyl site

Site on ribosome where growing polypeptide is attached.

804. polycistronic mRNA

Messenger RNA that makes more than one polypeptide on translation.

805. polyribosome

Configuration in which two or more ribosomes are translating the same mRNA molecule.

806. posttranscriptional control

Regulatory mechanisms that act on RNA transcripts already in cell.

807. posttranscriptional processing

Alteration of an RNA molecule to produce final RNA product.

808. posttranslational control

Regulatory mechanisms that act on polypeptides already made.

809. Pribnow box

Six nucleotide sequence in promoter of bacterial operon that determines first nucleotide to be used in transcription.

810. RNA splicing

Removal of introns from RNA to produce functional RNA molecule.

811. RNA-DNA hybridization

Association of complementary sequences of single-stranded DNA and RNA.

812. sigma factor

Unit of bacterial RNA polymerase that assures RNA synthesis will begin at correct site on DNA.

813.   spacer DNA

DNA segments of eukaryotic chromosomes not associated with nucleosomes and digested by deoxyribonuclease.

814.   stop codon

Nucleotide sequence in mRNA that signals the end of a message.

815.   substrate induction

Regulatory mechanism by which substrate activates enzymes that metabolize it.

816.   suppressor strain

Bacterial strain with tRNA that inserts an amino acid at the normal site of a stop codon.

817.   TATA box

Sequence of four nucleotides upstream from transcription initiation site that helps position RNA polymerase II.

818.   template

Pattern.

819.   termination signal

Point on a chromosome that indicates the end of a transcriptional unit.

820.   transcription

The transfer of coded genetic information from DNA to mRNA.

821.   transcription unit

DNA transcribed to a single, continuous RNA molecule.

822.   transcriptional control

Regulatory processes that control accessibility of chromosome regions and events at transcription site.

823.   transfer RNA

RNA that carries amino acids and places them in specific sites in a growing peptide chain.

824.   translation

The process by which mRNA codons are used to determine the sequence of amino acids in a protein.

825.   translational control

Regulation of translation by selection of mRNAs and changes in rates of mRNA degradation.

826.   translocation

Movement of a ribosome along an mRNA molecule so it can bind the next aminoacyl tRNA and elongate polypeptide chain.

827.   triplet code

Code that uses three units of information together.

828.   wobble

Flexibility in binding of codon and anticodon.

Microbes and Microbial Genetics

829.   adenovirus

Virus containing DNA and responsible for respiratory infections.

830.   amoeba

Unicellular organism that typically moves and feeds by pseudopodia.

831.   bacteriophage

A virus that invades bacterial cells.

832.   bacterium

A unicellular prokaryote lacking membrane-bound organelles.

833.   blue-green alga

Unicellular prokaryote lacking organelles but able to photosynthesize.

834.   colicin

Bacterial substance that can carry plasmids.

835.   conjugation

Mating and exchange of genetic information between bacteria.

836.   cryptic plasmid

Plasmid with no known function.

837.   diatom

Alga with a shell containing silicon.

838.   DNA tumor virus

Virus that contains DNA and causes tumors.

839.   fertility factor

Plasmid that confers maleness in some bacterial cells when transferred during sexual conjugation.

840.   fungus

Eukaryote that produces spores and obtains nutrients from other organisms.

841.   genophore

Genetic material of prokaryotes and viruses made of DNA or RNA not associated with proteins.

842.   ghost

Remains of bacteriophage particle on cell surface after phage's genetic information has entered the cell.

843.   herpes virus

Large virus containing DNA and causative agent for several diseases.

844.   Hfr bacterium

F-plus bacterium with F factor integrated into chromosome so it can be transferred during conjugation.

845.   i gene

Regulator gene in lac operon; DNA sequence that codes for repressor of lac operon.

846.   lactose operon

Bacterial genes that coordinate the induction of enzymes to metabolize lactose.

847.   lytic growth

Bacteriophage growth with many progeny phages and lysis of host cell.

848.   mating bridge

Structure across which DNA is transferred from male to female bacterium.

849.   oncogenic virus

Virus that can cause cancer.

850. operator

Nucleotide sequence in an operon that is recognized by a repressor molecule.

851. operon

A group of bacterial genes that function together and are controlled by a regulator gene.

852. origin of transfer

Point on plasmid at which transfer from male to female bacterium begins.

853. papovirus

Small DNA virus.

854. pilus

Delicate process from male bacterial cell across which DNA is transferred during conjugation.

855. plaque

Region in plate of bacterial cells where cells are infected with bacteriophage.

856. plasmid

Extrachromosomal DNA that replicates independently in a host cell.

857. promoter

DNA sequence to which RNA polymerase binds to start transcription; chemical that causes rapid division of mutant.

858. prophage

Bacteriophage DNA that has entered bacterial DNA.

859. protozoan

Unicellular or colonial, usually motile eukaryotic organism.

860. regulator gene

Nucleotide sequence that codes for a repressor, which regulates a prokaryotic operon.

861.   repression

Prevention of further enzyme synthesis by accumulation of end product.

862.   repressor

Protein that binds to operator site of operon and prevents transcription of adjacent structural genes.

863.   resistance factor

Set of genes for drug resistance found in certain plasmids.

864.   retrovirus

RNA virus that contains reverse transcriptase that allows it to use host double-stranded DNA for propagation.

865.   reverse transcriptase

An enzyme that makes DNA according to an RNA template.

866.   temperate bacteriophage

Bacteriophage that can either grow and lyse host cell or become integrated into host cells DNA.

867.   transducing bacteriophage

Bacteriophage that can incorporate host DNA into its genome and reversibly incorporate it in another host cell.

868.   transformation

Genetic modification by incorporation of exogenous DNA.

869.   tryptophan operon

Adjacent structural genes that code for enzyme needed to synthesize tryptophan and regulate its synthesis.

870.   virulent bacteriophage

Bacteriophage that always grows and lyses host cell.

871.   virus

Subcellular entity consisting of nucleotide and protein and incapable of replication outside a cell.

Reproduction

872. **acrosomal process**

Narrow channel in egg surface made of polymerized actin.

873. **acrosomal reaction**

Release of enzymes from acrosomal vesicle when a sperm contacts an egg.

874. **acrosomal vesicle**

Vesicle in the head of a sperm that contains enzymes that digest the surface coat of an egg.

875. **alternation of generations**

Plant life cycle in which haploid and diploid forms follow each other.

876. **asexual reproduction**

Kind of reproduction in which a single parent provides genetic information to offspring.

877. **bindin**

A sea urchin sperm protein that enables sperm to bind to egg membrane.

878. **cortical reaction**

Release of proteases from cortical vesicles that harden vitelline membrane and prevent further sperm penetration.

879. **cortical vesicle**

Membrane-bound sac in egg cell cortex that releases proteases when intracellular calcium concentration rises.

880. **egg**

Female haploid gamete.

881. **enucleation**

Removing of the nucleus from a eukaryotic cell.

882. **estrogen**

Steroid hormone produced by ovaries that affect female characteristics.

883. **external fertilization**

Union of gametes outside female body.

884. **fertilization**

Union of haploid gametes to form diploid zygote.

885. **gonad**

Reproductive organ where gametes are made.

886. **hormone**

Chemical substance that travels in blood and regulates function of target cells.

887. **internal fertilization**

Union of gametes within the female body.

888. **jelly coat**

Matrix of glycoproteins and polysaccharides around an egg, especially that of an amphibian.

889. **oogenesis**

Production of oocyte.

890. **oogonium**

Germ cell that will give rise to ovum.

891. **ootid**

Cell resulting from division of secondary oocyte that gives rise to ovum.

892. **ovary**

Female organ that produces gametes.

893.   ovulation

Release of ovum from ovary.

894.   ovum

Female haploid gamete; egg.

895.   polar body

Cell containing DNA and little cytoplasm produced during oogenesis.

896.   pollen grain

Male gametophyte enclosed in a tough coat produced by anther of angiosperm flower.

897.   pollen tube

Tube extending down the style of a carpel that delivers male gametes to ovule in angiosperm flower.

898.   pollination

Process by which pollen grains reach stigma in angiosperm flower.

899.   polyspermy

Fertilization of an ovum by more than one sperm.

900.   primary follicle

Cluster of cells around primary oocyte.

901.   primary germ layer

Any of three animal embryo tissue layers, ectoderm, mesoderm, and endoderm.

902.   primary oocyte

Cell derived from oogonium by mitosis that produces ovum by meiosis.

903.   primary spermatocyte

Cell derived from spermatogonium my mitosis that produces sperm by meiosis.

904. primordial germ cell

Cell that goes to genital ridge of embryo and gives rise to oogonia or spermatogonia depending on hormones released.

905. progesterone

Female steroid hormone that maintains pregnancy.

906. protamine

Small positively charged protein associated with DNA of sperm chromosomes.

907. pseudoplasmodium

Life cycle stage of a slime mold in which cells have aggregated.

908. pseudopod

Cell extension formed by cytoplasmic streaming.

909. reproduction

Process by which offspring arise.

910. secondary oocyte

Cell from first meiotic division of primary oocyte.

911. secondary spermatocyte

Cell from first meiotic division of primary spermatocyte.

912. seminiferous tubule

Site in testes where spermatogenesis occurs.

913. sexual reproduction

Reproduction requiring two parents each contributing to genetic information in offspring by gamete fusion.

914. somatic cell

Cell in multicellular organism not involved in reproduction of organism.

915. **spermatogenesis**

Production of sperm.

916. **spermatogonium**

Germ cell that eventually gives rise to sperm.

917. **spermatozoa**

Haploid gamete; sperm.

918. **terminal cell**

Cell from first division of plant zygote that gives rise to embryo.

919. **testis**

Male gonad.

920. **testosterone**

Male steroid hormone.

921. **vitelline membrane**

Membrane that surrounds an egg.

922. **zona pellucida**

Outer layer of mammalian egg which only a single sperm can penetrate.

923. **zygote**

Diploid cell produced by union of two haploid gametes.

Development

924. **animal-vegetal axis**

A polarity of an animal egg that indicates the embyro's orientation.

925. **axis of polarity**

Gradient within an embryo along which information for development exists.

926. **basal cell**

First progeny of a plant zygote that gives rise to suspensor tissue that holds an embryo to the ovule.

927. **blastocoel**

Fluid-filled cavity in the blastula of an animal embryo.

928. **blastomere**

One of many cells in the blastula of an animal embryo.

929. **blastula**

Stage of embryonic animal development consisting of a hollow ball of cells.

930. **cytoplasmic determinant**

Substance in egg cytoplasm that influences development.

931. **determination**

Commitment of a cell to a particular differentiation pathway.

932. **developmental potential**

Range of possible outcomes for a given cell in a developing organism.

933. **differentiation**

Developmental process in which cells become more specialized in structure and function.

934. **ectoderm**

Primary germ layer that gives rise to neural and epidermal tissues.

935. **embryogenesis**

Cell division and differentiation from a zygote to a multi-cellular organism.

936. **endoderm**

Primary germ layer that gives rise to the digestive tract lining of an embryo.

937. **fate map**

Diagram of an egg showing the developmental fate of each region.

938. **gastrulation**

Cell movements during embryonic development in which outer cells invaginate into the interior and create germ layers.

939. **germ cell**

Cell that produces haploid gametes.

940. **germ line**

Group of cells that give rise to gametes.

941. **gray crescent**

Surface of zygote where invagination occurs at gastrulation of amphibian embryo.

942. **growth**

Ability to increase in size.

943. **homeo box**

DNA sequence found in gene clusters responsible for pattern formation during development.

944. **homeotic mutation**

Change in DNA that causes one body structure to be replaced by another during development.

945. induction

Process in an embryo whereby one kind of cell or substance stimulates development of another kind of cell.

946. meristem

Part of of root or shoot with unlimited growth potential.

947. mesenchyme

Loose meshwork of cells from mesoderm of embryo that gives rise to muscle, bone, and connective tissue.

948. mesoderm

Primary germ layer that gives rise to muscle, bone, connective tissue, gonads, and blood.

949. metamorphosis

Dramatic morphological changes that change an organism from one body organization to another.

950. morphogen

Substance from one group of cells that cause morphogenic changes in another group of cells.

951. morphogenesis

Body pattern development during embryogenesis.

952. neural crest cell

Cell that has migrated from neural tube where it gives rise to nonneural tissues.

953. neural groove

Invagination in dorsal ectoderm induced by underlying mesoderm.

954. neural tube

Tube on dorsal surface of embryo formed from pinching off of neural groove.

955. neurulation

Process by which dorsal mesoderm induces formation of the neural tube.

956. **nuclear equivalence**

Idea that nuclei of multicellular organisms are genetically identical and differentiate by selective expression.

957. **organogenesis**

Formation of organs during development.

958. **pattern formation**

Production of a particular multicellular organization during organogenesis.

959. **pluripotent**

Cell or component capable of differentiating into different sets of progeny depending on environment.

960. **polar lobe**

Cytoplasm extending from a blastomere or zygote that lacks a nucleus.

961. **positional information**

Determination of nature of cell by its position in developing cells with respect to several axes.

962. **stem cell**

Animal cell capable of producing many different kinds of cells and tissues.

963. **totipotent**

Property of nucleus that retains ability to direct complete development of an embryo.

Evolution

964.   abiogenesis

Beginning without life; spontaneous generation.

965.   biogenesis

Generating life from life.

966.   biological evolution

Changes over time in living organisms.

967.   chemical evolution

The gradual increase in complexity of molecules thought to have preceded the origin of living cells.

968.   coacervate droplet

A mixture of large molecules thought to have preceded the organization of the first cells.

969.   common ancestor

An ancestor to two or more branches in the evolutionary tree.

970.   continuity of life

The transfer of life from parent to offspring.

971.   evolution

The process of change over time.

972.   fossil

Any evidence of organisms that lived in the past.

973.   natural selection

A process by which well adapted organisms survive and reproduce in larger numbers than less well adapted ones.

974.   oxidizing atmosphere

An atmosphere containing oxygen and other oxidizing molecules.

975.   primitive atmosphere

Atmosphere having gases that were present prior to the emergence of life on earth.

Kinds of Nutrition

976. **absorption**

Movement of substances across a membrane.

977. **absorptive**

Concerning absorption.

978. **autotroph**

An organism that makes organic compounds from inorganic substances in the environment.

979. **chemolithotroph**

Organisms that derive energy from inorganic compounds.

980. **chemosynthesis**

A process by which large molecules are made from small ones using energy from other chemicals.

981. **chemotroph**

An organism that gets energy from oxidizing inorganic or organic matter.

982. **digestion**

Breakdown of large molecules into smaller ones.

983. **facultative organism**

Organism that can function as aerobe or anaerobe.

984. **heterotroph**

An organism that metabolizes ready made organic matter.

985. **ingestion**

Intake of food or fluid.

986. **nutrition**

The act of providing substances needed for good health through food ingestion.

987. **photoautotroph**

Organism that uses carbon dioxide as a carbon source and sunlight as an energy source.

988. **photoauxotroph**

An organism that uses light energy to manufacture its own food.

989. **photoheterotroph**

Organism that can use energy from the sun but must have organic source of carbon.

990. **photolithotroph**

An organism that uses light energy to synthesize food from inorganic substances.

991. **phototroph**

Organism that uses light for energy.

Nerve Cells

992. **absolute refractory period**

Time during which sodium channel of a neuron is unable to respond to further depolarization.

993. **acetylcholine**

A neurotransmitter released by many axons, especially those that control skeletal muscles.

994. **action potential**

Wave of change in electrical potential across the cell membrane of an excited cell; impulse.

995. **adrenergic**

Concerning a neuron that releases norepinephrine (adrenalin).

996. **after-hyperpolarization**

Decrease in membrane potential below resting potential; occurs after action potential; undershoot.

997. **autonomic nervous system**

A part of the nervous system that regulates most visceral functions.

998. **axon**

The part of a neuron that typically carries impulses away from the cell body toward another neuron.

999. **axon terminal**

The end of an axon from which neurotransmitter is released.

1000. **axoplasm**

Cytoplasm inside the axon of a neuron.

1001. **catecholamine**

A class of amines that act as chemical messengers; dopamine, epinephrine, and norepinephrine.

1002. cell body

Part of a neuron where nucleus and organelles are located.

1003. central nervous system

Brain and spinal cord.

1004. chemical synapse

Junction between neurons across which chemicals diffuse.

1005. cholinergic

Relating to a neuron whose terminals release acetylcholine.

1006. cholinesterase

An enzyme that degrades acetylcholine.

1007. cholinesterase inhibitor

A substance that blocks cholinesterase action.

1008. dendrite

A cytoplasmic process of a neuron that commonly receives signals from other neurons.

1009. electrical synapse

Junction between two neurons across which impulses are transmitted by electrical signals.

1010. endoplasm

Fluid cytoplasm in the interior of a cell capable of amoeboid movement.

1011. epinephrine

Hormone from the adrenal medulla that regulates metabolic responses to stress.

1012. equilibrium potential

Potential across a membrane exactly equal to the effect of the concentration gradient of a given kind of ion.

1013. excitatory postsynaptic potential (EPSP)

Change toward excitation of a membrane potential in response to neurotransmitter binding to receptors.

1014. Goldman equation

Modification of Nernst equation for calculating the resting potential of a membrane by summing ion effects.

1015. inhibitory postsynaptic potential (IPSP)

Change away from excitation of a membrane potential in response to neurotransmitter binding to receptors.

1016. interneuron

Neuron that receives signals from sensory neuron and relays them to motor neuron.

1017. membrane potential

Gradient of charge across a membrane.

1018. motor neuron

Neuron that transmits impulses to a muscle or gland.

1019. myelin

An insulating substance deposited around axons.

1020. myelin sheath

Concentric membrane layers that surround and insulate an axon.

1021. Nernst equation

Equation used to calculate equilibrium potential for a particular ion.

1022. nerve impulse

Action potential propagated along an axon.

1023. nervous system

Neurons and associated tissues that respond to stimuli and relay action potentials.

1024. **neurilemma**

The Schwann cell membrane.

1025. **neuron**

Cell specialized to conduct action potentials.

1026. **neurotoxin**

Substance that is toxic to neurons and disrupts action potentials.

1027. **neurotransmitter**

A chemical substance from one neuron that transmits a signal to another neuron at a synapse.

1028. **node of Ranvier**

A gap in an axon's myelin sheath.

1029. **noradrenalin**

Norepinephrine.

1030. **norepinephrine**

A neurotransmitter of the sympathetic division of the autonomic nervous system and of some brain neurons.

1031. **oligodendrocyte**

Cell of central nervous system that produces myelin.

1032. **opsin**

A protein that combines with retinine in the retina.

1033. **patch clamping**

Technique for studying membrane channels using micropipette sealed to cell surface for application of current.

1034. **perikaryon**

Substance around a nucleus; the cell body of a neuron.

1035. peripheral nervous system

Sensory and motor neurons that carry signals to and from the central nervous system.

1036. postsynaptic

Referring to a neuron that receives a neurotransmitter at a synapse.

1037. postsynaptic membrane

Membrane with receptors for neurotransmitter that sends action potentials to next synapse.

1038. presnyaptic membrane

Axon membrane that releases neurotransmitters into synaptic cleft.

1039. presynaptic

Referring to a neuron that releases a neurotransmitter at a synapse.

1040. process

Axon or dendrite of neuron.

1041. proteoliposome

Laboratory-made membrane containing specific proteins in a phospholipid bilayer.

1042. refractory period

Time between successive stimuli during which a neuron is unable to respond to a stimulus.

1043. resting potential

Potential difference across membrane of unstimulated neuron.

1044. retinene

A carotenoid pigment that binds to opsin.

1045. rhodopsin

A light-sensitive protein found in rods of the retina.

1046. Schwann cell

A myelin producing cell in the peripheral nervous system.

1047. sensory neuron

Cell capable of receiving information from the environment.

1048. somatic nervous system

Part of peripheral nervous system that receives sensory input and controls skeletal muscle activity.

1049. spatial summation

Addition of postsynaptic potentials from various sites along membrane of neuron.

1050. stimulus

An event that typically elicits a response.

1051. synapse

A junction where a signal passes from one neuron to the next in a pathway, usually by neurotransmitter diffusion.

1052. synaptic cleft

Space between presynaptic and postsynaptic neurons across which neurotransmitters diffuse.

1053. synaptic knob

Structure on end of axon that contains neurotransmitter vesicles.

1054. synaptic vesicle

Membrane-bound structure containing neurotransmitter.

1055. temporal summation

Addition of postsynaptic potentials reaching neuron membrane in rapid succession.

1056. threshold stimulus

Stimulus sufficiently strong to depolarize a membrane.

1057.   transducin

Enzyme involved in visual process.

1058.   transduction

Transport by a virus of DNA from one cell to another.

Muscle Cells and Movement

1059. **actin**

A contractile protein.

1060. **agonist**

One substance that mimics another substance.

1061. **amoeboid movement**

Movement by pseudopodia.

1062. **antagonist**

A substance that prevents another substance from acting.

1063. **atrophy**

A decrease in size, usually accompanied by reduced function.

1064. **bacterial flagellum**

A spiral filament attached to the membrane of a bacterium and used for locomotion.

1065. **basal body**

Microtubular anchor at the base of a flagellum or cilium; has same structure as a centriole.

1066. **calmodulin**

Regulatory calcium-binding protein in eukaryotic cells.

1067. **calsequestrin**

Calcium-binding protein of sarcoplasmic reticulum that increases calcium storage in its cisternae.

1068. **cardiac muscle**

Striated muscle with intercalated discs; makes up heart muscle.

1069.   chemotaxis

Detection of a chemical and movement toward or away from it.

1070.   contractile protein

A protein that acts in shortening a muscle or causing it to develop tension.

1071.   contractility

The ability to develop tension or shorten.

1072.   contraction cycle

Repetitive sliding actions of actin and myosin in a muscle filament as it develops tension.

1073.   creatine phosphate

A molecule that accounts for limited energy storage in muscle.

1074.   cross-bridge

The end of a myosin filament bound to actin during muscle contraction.

1075.   cyclosis

Circular flow of contents of a cell around the central vacuole.

1076.   cytochalasins

Drugs derived from fungi that inhibit actin polymerization.

1077.   cytoplasmic streaming

Back-and-forth flow of cytoplasm seen in plant and algal cells.

1078.   dynein

Protein that extends between microtubule doublets in an axoneme.

1079.   ectoplasm

Thick cytoplasm beneath the plasma membrane and capable of amoeboid movement.

1080. **F-actin**

Polymer of G-actin.

1081. **flagellin**

Protein unit found in spiral filaments of a bacterial flagellum.

1082. **G-actin**

Globular protein unit of actin.

1083. **heavy meromyosin**

Portion of myosin molecule that contains ATPase heads.

1084. **isometric**

Having the same length.

1085. **light meromyosin**

Polypeptide part of myosin molecule not containing ATPase.

1086. **microfilament-based movement**

Motility derived form actions of actin and myosin.

1087. **microtubule-based movement**

Motility derived from actions of cilia, flagella, and spindle fibers.

1088. **motility**

Ability to move.

1089. **muscle fiber**

Long, multinucleate cell differentiated for contraction.

1090. **myofibril**

Unit of muscle contraction consisting of bundles of thick and thin filaments.

1091. myofilament

A component of a myofibril consisting of one or more protein molecules.

1092. myoglobin

A pigmented protein that binds oxygen in muscle tissue.

1093. myokinase

Enzyme that makes ATP from two molecules of ADP.

1094. myosin

A protein that comprises thick filaments of a myofibril.

1095. phalloidin

Drug that blocks actin depolymerization and prevents cell movements.

1096. radial spoke

Inward projection from a microtubule doublet to center microtubules in axoneme that participates in bending.

1097. sarcomere

The contractile unit of skeletal muscle.

1098. sarcoplasm

The protoplasmic, nonfibrillar substance of a muscle cell.

1099. sarcoplasmic reticulum

A vesicular network associated with myofibrils of a striated muscle cell.

1100. shuttle streaming

Cytoplasmic flow that reverses direction with predictable periodicity.

1101. sliding filament theory

An explanation of how myofilaments move with respect to each other during muscle contraction.

1102. **sliding microtubule model**

Explanation of motility by sliding of outer doublets past each other causing localized bending.

1103. **striated muscle**

Skeletal muscle that carries out voluntary movements.

1104. **syncytium**

A group of cells that lack membranes to separate them.

1105. **thick filament**

Bundle of myosin molecules in striated muscle myofilament.

1106. **thin filament**

Bundle of actin filaments with tropomyosin and troponin attached in striated muscle myofilament.

1107. **transverse (T) tubule**

Crosswise tubule in skeletal muscle myofibrils that carries signals from the sarcolemma to the myofibrils.

1108. **tropomyosin**

A muscle protein that alters the actin configuration so that contraction can occur.

1109. **troponin**

A muscle protein that binds to tropomyosin causing it to alter the configuration of actin.

Immunology

1110. **acquired immunodeficiency syndrome (AIDS)**

Viral infection that destroys immune response of T cells.

1111. **activated T cell**

T cell that has been processed and made capable of initiating immune reactions.

1112. **antibody repertoire**

Total assortment of an organism's different antibodies that will allow it to recognize any foreign substances.

1113. **antigen**

A substance that can elicit an immune response.

1114. **antigen receptor**

Lymphocyte surface molecule that responds to a particular antigenic determinant.

1115. **antigen-binding site**

A variable region of an immunoglobulin that can bind a specific antigen.

1116. **antigen-dependent differentiation**

Stage in differentiation of a lymphocyte that begins when a particular antigen binds to its surface receptors.

1117. **antigen-independent differentiation**

Stage in differentiation of a lymphocyte in which uncommitted cell makes a specific antigen receptor.

1118. **antigenic determinant**

A particular part of an antigen to which the immune system responds.

1119. **autoimmune reaction**

Process by which the immune system attacks an organism's own tissues.

1120. B-lymphocyte

White blood cell capable of producing antibody and humoral response of the immune system.

1121. C gene

Genetic information for constant region of immunoglobulin.

1122. cell-mediated immune response

Immune reaction in which T lymphocytes lead direct attack on foreign substances.

1123. central lymphoid tissue

Site at which lymphocytes mature.

1124. clonal selection theory

Lymphocytes destined to act on a particular antigen can be triggered to produce large clones when sensitized.

1125. constant region

Nonvariable region of an immunoglobulin.

1126. cytotoxic T cell

T cell activated to lyse foreign cells.

1127. D segment

DNA that codes for the most variable portion of the heavy chain of an immunoglobulin.

1128. effector cell

A differentiated lymphocyte that carries out an activity of the immune system, plasma cell or activated T cell.

1129. fluorescent antibody

Antibody to which fluorescent dye molecules are bound and whose fluorescence allows it to be localized.

1130. graft rejection

Immune response to a tissue recognized as foreign.

1131. **helper T cell**

Lymphocyte that promotes immune response of other lymphocytes.

1132. **hematopoietic stem cell**

Cell that gives rise to blood cells, including lymphocytes.

1133. **hematopoietic tissue**

Tissue containing cells that give rise to blood cells.

1134. **HLA complex**

Set of genes that encode HLA antigens.

1135. **human leukocyte-associated antigen (HLA)**

Antigen on cell surface of human cells that is recognized by the immune system.

1136. **humoral antibody**

Immunoglobulin released by plasma cells that circulates in blood.

1137. **humoral response**

Release of immunoglobulins (antibodies) that recognize and bind to an antigen.

1138. **hybridoma**

Hybrid cells that produce monoclonal antibodies made by fusing antibody-producing B cell with myeloma cell.

1139. **immune response**

Process of recognizing a foreign substance and acting to remove it from the body or inactivate it.

1140. **immune system**

Diffuse vertebrate body system that defends against foreign substances.

1141. **immunity**

Resistance to a pathogen as a result of prior exposure.

1142. interleukin-1

Lymphokine from macrophage that stimulates T cells to secrete interleukin-2.

1143. interleukin-2

Lymphokine from T cells that binds to receptors of activated lymphocytes making them proliferate.

1144. lymphocyte

Kind of white blood cells that participates in immune functions.

1145. lymphoid tissue

Tissue that contains lymphocytes and participates in an immune reaction.

1146. lymphokine

Substance secreted by T lymphocyte that activates other lymphocytes; interleukin.

1147. macrophage

Large phagocytic cell.

1148. major histocompatibility complex (MHC)

Genes that code for cell-surface antigens, which give cells a unique identity.

1149. memory cell

Long-lasting lymphocyte sensitized to an antigen that will initiate an immune response if the antigen is encountered.

1150. monoclonal antibody

Like immunoglobulins derived from cells of a single clone.

1151. myeloma protein

Portion of antibody secreted in large amounts by malignant plasma cells.

1152. peripheral lymphoid tissue

Organs such as the spleen, lymph nodes, and tonsils that contain lymphocytes.

1153. plasma cell

Progeny of sensitized B lymphocyte capable of producing immunoglobulin that binds foreign substance.

1154. primary immune response

First immune response to a particular antigen.

1155. regulatory T cell

Lymphocytes that modify immune responses.

1156. secondary immune response

Response to a previously encountered antigen in which a larger, quicker response is possible.

1157. secretory component

Polypeptide from IgA receptor that remains associated with IgA molecule.

1158. T lymphocyte

Lymphocyte involved in cell-mediated immune responses and in modulating humoral responses.

1159. transplantation antigen

Cell-surface antigen in grafted tissue that produces immune response; histocompatibility antigen.

1160. variable domain

Portion of antibody molecule that specifically binds to a particular antigen.

Microscopy / Other Techniques

1161. **A-face**

Interior face of cytoplasmic membrane revealed by freeze-fracturing.

1162. **angular aperature**

A measure of the quantity of light entering the objective lens of a compound microscope.

1163. **autoradiography**

Procedure in which radioactive substances in a cell are located by their effects on photographic film.

1164. **B-face**

Interior face of the outer surface of a membrane made visible by freeze-fracturing.

1165. **birefringence**

Ability to rotate the plane of polarized light.

1166. **birefringent microscopy**

Light microscopy in which colored specimen causes different amounts of light to pass through so image can be formed.

1167. **compound microscope**

Light microscope that uses several lenses in combination.

1168. **condensor lens**

Lens that directs light or electrons toward a specimen.

1169. **cryoprotection**

Decreasing ice crystal formation during freezing of specimen for microscopy.

1170. **darkfield microscopy**

Technique that makes specimen visible by light diffracted from it.

1171.   deep etching

Ultra-rapid freeze etching that produces surfaces from
interior of cell.

1172.   differential interference contrast microscopy

Light microscopy that uses polarized light to create
illusion of three-dimensional image.

1173.   differential centrifugation

Technique used to separate organelles and cell components
according to density.

1174.   electron gun

Part of electron microscope that emits electrons that
generate the electron beam that is focused on a specimen.

1175.   electron micrograph

Photograph of a specimen exposed to a photographic plate and
an image-forming electron beam.

1176.   electron microscope

Instrument that uses an electron beam to create an image of
a specimen with greater magnification than with light.

1177.   fluorescence

Ability to absorb ultraviolet light and reemit energy as
visible light.

1178.   fluorescence microscopy

Light microscopy that produces an image on a dark background
by using ultraviolet light to make the specimen fluoresce.

1179.   fluorescent dye

Substance that emits visible light when it absorbs
ultraviolet light.

1180.   focal length

Distance between midline of a lens and point at which rays
passing through it converge to a focal point.

1181.   follicle cell

One of many cells around a mammalian egg that nourish it and
produce hormones.

1182. freeze-etching

Procedure for cleaving a quick-frozen specimen and etching that makes a small area of a cell surface visible.

1183. freeze-fracturing

Procedure for cleaving a quick-frozen specimen in a vacuum and shadowing to make a replica of fractured cell surface.

1184. high-voltage electron microscope

Electron microscope that uses accelerating voltages that allow study of thicker specimens.

1185. lens system

Arrangement of glass or electromagnetic lenses used to focus and enlarge a microscopic specimen.

1186. light microscope

Instrument that uses lenses and visible light to enlarge image of specimen.

1187. micrometer

Unit of measure equal to 1 millionth of a meter; a micron.

1188. microscope

An instrument for observing structures too small to see with the naked eye.

1189. microtome

Instrument for slicing tissue into thin sections for microscopic study.

1190. negative contrast

Optical system with lightly stained specimen seen against a dark background.

1191. negative staining

Technique in which stain is placed around specimen, especially in transmission electron microscopy.

1192. numerical aperature (NA)

A measure of the refractive index of the medium between a specimen and the objective lens of a light microscope.

1193. **objective lens**

Lens of a light or electron microscope found immediately above specimen and that forms primary image.

1194. **ocular lens**

Eyepiece of a compound light microscope that enlarges primary image.

1195. **optical sectioning**

Technique of focusing on different planes in a microscopic specimen.

1196. **perfusion**

Technique in which substance is passed through the blood stream of an animal.

1197. **phase-contrast microscopy**

Light microscopy that produces an image with highly contrasting light and dark areas by manipulating diffraction.

1198. **plane-polarized light**

Light in which waves are traveling in one plane.

1199. **polarization microscopy**

Technique of light microscopy that uses a polarizer and an analyzer to provide polarized light for viewing specimen.

1200. **poststaining**

Staining of a tissue with lead or uranium after preparation for electron microscopy.

1201. **primary fixation**

First step in preparing a specimen for microscopy; hardens and stabilizes components of cells.

1202. **projector lens**

Electron microscope lens that produces image on photographic plate of viewing screen.

1203. **refractive index**

Velocity of light as it passes from one medium to another relative to its velocity in a vacuum.

1204. **resolution**

Shortest distance between two objects that can be seen distinctly.

1205. **scanning electron microscope**

Kind of microscope in which electron beam is cast over specimen and image formed by electrons emitted from specimen.

1206. **scanning transmission electron microsc.**

Microscope in which electron beam is cast over specimen and image formed by electrons transmitted through specimen.

1207. **shadowing**

Use of layer of electron-dense material on specimen surface toward the electrode.

1208. **stage**

Platform for specimen on light microscope.

1209. **staining**

Procedure that gives specimen color or electron density for microscopic viewing.

1210. **stereo pair**

Two photographs of same object from slightly different angles that give 3-D impression viewed stereoscopically.

1211. **thin section**

Thin slice of specimen on glass slide for viewing with light microscope.

1212. **transmission electron microscope (TEM)**

Kind of electron microscope that forms an image by different degree to which specimen parts transmit electrons.

1213. **ultracentrifuge**

Instrument that separates organelles and macromolecules by movement during centrifugation at very high speed.

1214. **ultramicrotome**

Instrument that can slice specimens into ultrathin sections for electron microscopy.

1215. **vacuum evaporator**

Jar containing electrodes in which a vacuum can be created for preparing metal replicas of biological surfaces.

1216. **wavelength**

Distance between crests of successive waves.

Index

A-face	121	amphipathic molecule	16
abiogenesis	100	amylase	20
absolute refractory period	104	amylopectin	9
absorption	102	amyloplast	25
absorption spectrum	55	amylose	9
absorptive	102	anabolic	45
accessory pigment	55	anabolic pathway	45
acetyl coenzyme A	50	anabolism	45
acetylcholine	104	anaerobic	50
acid	5	anaphase	69
acquired immunodeficiency		anaphase I	69
syndrome (AIDS)	116	anastral mitosis	69
acrocentric chromosome	69	Angstrom	1
acrosomal process	91	angular aperature	121
acrosomal reaction	91	animal-vegetal axis	96
acrosomal vesicle	91	anion	5
actin	111	annulus	34
actinomycin D	60	antagonist	111
action potential	104	antibody repertoire	116
action spectrum	55	anticodon	81
activated monomer	34	antigen	116
activated T cell	116	antigen receptor	116
activation	39	antigen-binding site	116
activation energy	20	antigen-dependent differentiation	116
active site	20	antigen-independent	
active transport	39	differentiation	116
adaptive enzyme synthesis	81	antigenic determinant	116
adenine	60	antimitotic drug	69
adenosine	60	asexual reproduction	91
adenosine diphosphate (ADP)	50	aster	69
adenosine monophosphate (AMP)	50	astral mitosis	69
adenosine triphosphate	45	asymmetric carbon atom	9
adenovirus	87	atom	5
adrenergic	104	atomic number	5
adsorptive endocytosis	39	atomic weight	5
aerobic	50	ATP synthetase complex	50
after-hyperpolarization	104	atrophy	111
agonist	111	attenuation	81
alcoholic fermentation	50	attenuation site	81
alkaline	5	autoimmune reaction	116
allele	75	autolysis	20
allosteric effector	20	autonomic nervous system	104
allosteric enzyme	20	autophagic lysosome	34
allosteric regulation	20	autophagic vacuole	34
alpha helix	16	autophagy	39
alternation of generations	91	autoradiography	121
Ames test	75	autotroph	102
amide bond	20	axis of polarity	96
amino acid	16	axon	104
aminoacyl site	81	axon terminal	104
aminoacyl tRNA	81	axoneme	25
aminoacyl tRNA synthetase	81	axoplasm	104
aminopeptidase	20	B-DNA	60
amoeba	87	B-face	121
amoeboid movement	111	B-lymphocyte	117
amphibolic pathway	45	backcross	75

bacterial flagellum	111	cell body	105
bacteriochlorophyll	55	cell coat	25
bacteriophage	87	cell cycle	69
bacteriorhodopsin	34	cell junction	34
bacterium	87	cell membrane	34
basal body	111	cell plate	69
basal cell	96	cell theory	1
base	5	cell wall	25
base pairing	60	cell-mediated immune response	117
belt desmosome	34	cellular secretion	39
benign tumor	75	cellular transport	39
beta oxidation	50	cellulose	9
beta pleated sheet	16	central dogma	81
bidirectional replication	66	central granule	32
bile acid	13	central lymphoid tissue	117
bindin	91	central nervous system	105
binding site	34	central vacuole	25
biochemistry	1	centriole	25
bioenergetics	45	centromere	69
biogenesis	100	ceramide	13
biological evolution	100	cerebroside	13
biology	1	CF-one	55
bioluminescence	45	CF-zero	56
biopolymer	9	Chargaff's rules	60
biosphere	1	charge repulsion	5
biosynthesis	20	chemical evolution	100
biotechnology	1	chemical synapse	105
birefringence	121	chemisomotic model	50
birefringent microscopy	121	chemolithotroph	102
bivalent	75	chemosynthesis	102
blastocoel	96	chemotaxis	112
blastomere	96	chemotroph	102
blastula	96	chiasma	70
blue-green alga	87	chitin	9
bond energy	21	chlorophyll	56
buffer	5	chloroplast	25
bundle sheath cell	55	cholesterol	13
buoyant density	39	cholinergic	105
C gene	117	cholinesterase	105
C-four plant	55	cholinesterase inhibitor	105
C-three plant	55	chromatid	70
calmodulin	111	chromatin	32
calorie	21	chromatin fiber	32
calsequestrin	111	chromoplast	25
Calvin cycle	55	chromosomal fiber	70
cancer	75	chromosome	32
carbohydrate	9	chromosome mapping	75
carboxypeptidase	21	chromosome puff	32
carcinogen	75	chromosome theory of heredity	75
carcinoma	75	cilium	25
cardiac muscle	111	cisterna	25
carotenoid	55	cisternal space	26
catabolic activator protein (CAP)	81	cistron	82
catabolic pathway	45	citric acid cycle	50
catabolism	45	clathrin	34
catabolite repression	81	cleavage	70
catalyst	21	cleavage furrow	70
catecholamine	104	clonal selection theory	117
cation	5	clone	76
cell	1	cloning	82

closed system	45	cytology	1
coacervate droplet	100	cytoplasm	1
coated pit	35	cytoplasmic determinant	96
coated vesicle	39	cytoplasmic streaming	112
coding strand	82	cytosine	61
codon	82	cytoskeleton	26
coenzyme	21	cytosol	26
coenzyme A (CoA)	51	cytotoxic T cell	117
coenzyme Q (CoQ)	51	D segment	117
cohesive end	60	dark reaction	56
colchicine	70	darkfield microscopy	121
colicin	87	data	2
colinear	16	daughter cell	70
collagen	16	deamination	16
colloid	6	deductive reasoning	2
colloidal dispersion	6	deep etching	122
common ancestor	100	degenerate	82
compartmentalization	1	dehydration	21
competitive inhibition	21	dehydrogenation	51
complementary base pairing	82	denaturation	16
compound	6	dendrite	105
compound microscope	121	density gradient	40
concentration work	46	deoxyribonuclease	61
condensation reaction	21	deoxyribonucleic acid (DNA)	61
condensing vacuole	26	deoxyribose	61
condensor lens	121	depurination	61
conjugation	87	desmosome	35
connexon	39	desmsotubule	35
consensus sequence	60	determination	96
constant region	117	development	2
constitutive heterochromatin	60	developmental potential	96
contact inhibition	70	diakineses	70
continuity of life	100	diatom	87
contractile protein	112	dictyosome	26
contractile ring	70	differential interference contrast	
contractility	112	microscopy	122
contraction cycle	112	differential centrifugation	122
core glycosylation	82	differentiation	96
core particle	60	diffusion	40
Cori cycle	51	digestion	102
cortical cytoplasm	26	digitalis	40
cortical reaction	91	diglyceride	13
cortical vesicle	91	diploid	76
cotranslational transport	82	diplotene	71
coupled transport	39	direct mutagen	76
covalent bond	46	disaccharide	9
creatine phosphate	112	disulfide bond	46
crista	26	division phase	71
cross-bridge	112	DNA glycosidase	61
crossing-over	76	DNA gyrase	66
cryoprotection	121	DNA ligase	66
cryptic plasmid	87	DNA polymerase	66
cyclic AMP (cAMP)	51	DNA replication	66
cyclic electron flow	56	DNA tumor virus	87
cyclic photophosphorylation	56	docking protein	82
cyclosis	112	dominant	76
cytochalasins	112	double bond	46
cytochrome	51	double helix	61
cytochrome C oxidase	51	drug detoxification	40
cytokinesis	70	dynein	112

ectoderm	97	facilitated transport	40
ectoplasm	112	facultative heterochromatin	61
effector	82	facultative organism	102
effector cell	117	fate map	97
effector site	21	fatty acid	13
egg	91	feedback	2
electrical synapse	105	feedback inhibition	83
electrical work	46	fermentation	52
electrochemical gradient	35	fertility factor	88
electron	6	fertilization	92
electron gun	122	fibrillar component	83
electron micrograph	122	fibrous protein	17
electron microscope	122	filtration	40
electron transport	51	first law of thermodynamics	46
electron transport system	51	Fischer projection	9
element	6	flagellin	113
embryogenesis	97	flagellum	26
Emerson enhancement effect	56	flavin adenine dinucleotide (FAD)	52
end-product repression	82	flavin mononucleotide (FMN)	52
endergonic	46	flavoprotein	52
endocytosis	40	fluid-mosaic model	35
endoderm	97	fluidity	35
endopeptidase	21	fluorescence	122
endoplasm	105	fluorescence microscopy	122
endoplasmic reticulum	26	fluorescent antibody	117
endoreplication	66	fluorescent dye	122
endosome	35	focal length	122
energy	46	follicle cell	122
entropy	46	food cup	35
enucleation	91	forming face	26
environment	2	fossil	100
enzyme	21	frameshift mutation	76
enzyme kinetics	22	free energy	47
epinephrine	105	free energy change	47
equilibrium constant	22	freeze-etching	123
equilibrium potential	105	freeze-fracturing	123
essential amino acid	16	functional group	9
essential fatty acid	13	fungus	88
ester bond formation	46	G-1 phase	71
estrogen	92	G-2 phase	71
euchromatin	61	G-actin	113
eukaryotic	2	gamete	71
evolution	100	gametic meiosis	71
excitability	2	gametogenesis	71
excitatory postsynaptic potential		gametophyte	71
(EPSP)	106	gap junction	35
exergonic	46	gastrulation	97
exocytosis	40	gel	26
exon	61	gene	76
exopeptidase	22	gene amplification	76
extensin	16	generation time	71
external fertilization	92	genetic code	76
extracellular	40	genetics	2
extracellular digestion	40	genome	77
F-actin	113	genomic control	77
F-one generation	76	genophore	88
F-one particle	51	genotype	77
F-two generation	76	germ cell	97
F-zero particle	51	germ line	97
facilitated diffusion	40	ghost	88

globular protein	17
gluconeogenesis	52
glucose	10
glycerol	13
glycine	17
glycocalyx	35
glycogen	10
glycogenesis	52
glycogenolysis	52
glycolate pathway	56
glycolipid	13
glycolysis	52
glycoprotein	17
glycosaminoglycan	35
glycosidic bond	47
glycosphingolipid	13
glycosylation	83
glyoxylate cycle	52
glyoxysome	27
Goldman equation	106
Golgi apparatus	27
Golgi-associated ER	27
gonad	92
gradient	41
graft rejection	117
gram molecular weight	6
granular component	83
granum	56
gray crescent	97
growth	97
growth phase	71
guanine	61
gyrase	66
half-life	2
haploid	77
Haworth projection	10
heat	47
heat of vaporization	47
heavy meromyosin	113
helicase	66
helix	62
helper T cell	118
hematopoietic stem cell	118
hematopoietic tissue	118
heme	17
hemicellulose	10
hemoglobin	17
heptose	10
herpes virus	88
heterochromatin	77
heterogeneous nuclear RNA	66
heterophagic lysosome	36
heterotroph	102
heterozygous	77
hexose	10
Hfr bacterium	88
high-voltage electron microscope	123
histone	62
HLA complex	118
holoenzyme	22
homeo box	97
homeopolymer	10
homeotic mutation	97
homozygous	77
hormone	92
human leukocyte-associated antigen (HLA)	118
humoral antibody	118
humoral response	118
hybrid	77
hybridoma	118
hydrogen bond	47
hydrogenation	52
hydrolase	22
hydrolysis	6
hydrophilic	14
hydrophobic	14
hydrophobic interactions	14
hydrostatic pressure	41
hydroxylation reaction	22
hyperosmotic	41
hypertonic	41
hyposmotic	41
hypothesis	2
hypotonic	41
i gene	88
immune response	118
immune system	118
immunity	118
in vitro	2
in vivo	3
indirect mutagen	77
induced-fit model	22
inducible enzyme	22
induction	98
inductive reasoning	3
ingestion	102
inhibition	22
inhibitory postsynaptic potential (IPSP)	106
initiation codon	83
inner membrane	27
inner nuclear membrane	32
insertional mutagenesis	62
integral membrane protein	17
interleukin-1	119
interleukin-2	119
intermediate filament	27
intermembrane space	27
internal fertilization	92
interneuron	106
interphase	71
intracellular	41
intracellular transport	41
intrathylakoid space	56
intron	83
ion	6
ion carrier	36
ionic bond	47
ionophore	36

iron-sulfur protein	17	major histocompatibility complex	
irreversible inhibitor	22	(MHC)	119
isomer	10	malignancy	78
isometric	113	maltose	11
isosmotic	41	map unit	78
isothermal	47	mass	3
isotonic	41	mating bridge	88
isotope	6	matrix	28
jelly coat	92	maturing face	28
karyotype	72	maximum velocity	23
keratin filament	27	mechanical work	47
ketosugar	10	median	3
kilocalorie	47	meiosis	72
kinase	22	membrane	36
kinetic	6	membrane asymmetry	36
kinetochore	32	membrane potential	106
lactate fermentation	52	membrane turnover	36
lactose	10	memory cell	119
lactose operon	88	Mendelian genetics	78
lagging strand	66	meristem	98
lampbrush chromosome	32	mesenchyme	98
large ribosomal unit	27	mesoderm	98
lateral diffusion	41	mesophyll cell	56
law of independent assortment	77	messenger RNA	62
law of segregation	77	metabolic pathway	47
leader peptide	83	metabolism	48
leader sequence	83	metabolite	48
leading strand	67	metacentric chromosome	72
leaf peroxisome	27	metamorphosis	98
lecithin	14	metaphase	72
lectin	17	metaphase I	72
lens system	123	metastasis	78
leptotene	72	meter	3
lethal synthesis	53	Michaelis constant	23
ligand	36	Michaelis-Menten equation	23
light meromyosin	113	microbody	28
light microscope	123	microfibril	28
light reaction	56	microfilament	28
lignin	10	microfilament-based movement	113
Lineweaver-Burk equation	23	micromere	28
linkage group	78	micrometer	123
lipid	14	micronutrient	53
lipid bilayer	36	micropinocytosis	42
lipid body	27	microscope	123
lipid monolayer	36	microsome	28
lipoprotein	14	microtome	123
liter	3	microtrabecular lattice	28
lock-and-key model	23	microtubule	28
lumen	3	microtubule-associated protein	
lymphocyte	119	(MAP)	28
lymphoid tissue	119	microtubule-based movement	113
lymphokine	119	microvillus	28
lysosome	27	mineral	53
lysozyme	23	minor groove	62
lytic growth	88	mitochondrion	29
macromolecule	11	mitogen	78
macronutrient	53	mitosis	72
macrophage	119	mitotic index	72
macropinocytosis	42	mitotic spindle	72
major groove	62	mixture	6

mole	7	nonhistone chromosomal protein	62
molecule	7	nonpolar	14
monocistronic mRNA	83	nonreiterated sequence	62
monoclonal antibody	119	nonsense mutation	78
monoglyceride	14	noradrenalin	107
monolayer of cells	72	norepinephrine	107
monomer	11	nuclear	32
monomeric protein	17	nuclear cortex	32
monosaccharide	11	nuclear envelope	33
morphogen	98	nuclear equivalence	99
morphogenesis	98	nuclear matrix filament	33
motility	113	nuclear organizer region (NOR)	33
motor neuron	106	nuclear pore	33
mucopolysaccharide	11	nucleic acid	62
mucoprotein	17	nucleoid	33
multimeric protein	18	nucleolus	33
muscle fiber	113	nucleoplasm	33
mutagen	78	nucleoside	62
mutation	78	nucleosome	63
myelin	106	nucleotide	63
myelin sheath	106	nucleus	7
myeloma protein	119	numerical aperature (NA)	123
myofibril	113	nutrition	103
myofilament	114	objective lens	124
myoglobin	114	obligate aerobe	53
myokinase	114	obligate anaerobe	53
myosin	114	ocular lens	124
N-formyl methionine	83	Okazaki fragment	67
NADH dehydrogenase	53	oligodendrocyte	107
nanometer	3	oncogene	78
nascent polypeptide	83	oncogenic virus	88
natural selection	100	oogenesis	92
negative contrast	123	oogonium	92
negative control	84	ootid	92
negative regulation	48	open system	48
negative staining	123	operator	89
negative supercoil	62	operon	89
Nernst equation	106	opsin	107
nerve impulse	106	optical sectioning	124
nervous system	106	organelle	29
neural crest cell	98	organic	7
neural groove	98	organogenesis	99
neural tube	98	origin of replication	67
neurilemma	107	origin of transfer	89
neurofilament	29	osmolarity	42
neuron	107	osmosis	42
neurotoxin	107	osmotic pressure	42
neurotransmitter	107	oubaine	42
neurulation	98	outer membrane	29
neutron	7	outer nuclear membrane	33
nicotinamide adenine		ovary	92
dinucleotide (NAD)	53	ovulation	93
nicotinamide adenine		ovum	93
dinucleotide phosp.	57	oxidation	7
node of Ranvier	107	oxidation-reduction couple	53
noncellulosic matrix	11	oxidative deamination	18
noncoding strand	67	oxidative phosphorylation	53
noncompetitive inhibition	23	oxidizing atmosphere	101
noncyclic electron flow	57	oxygenic photoautotroph	57
noncyclic photophosphorylation	57	P/O ratio	57

pachytene	72	pinocytosis	42
packing ratio	63	plane-polarized light	124
palindromic	63	plaque	89
palisade cell	29	plasma cell	120
papovirus	89	plasma membrane	37
passive transport	42	plasmid	89
patch clamping	107	plasmodesma	29
pattern formation	99	plasmolysis	42
pectin	11	plastid	29
pentose	11	plastocyanin	58
peptidase	23	plastoquinone	58
peptide bond	48	pluripotent	99
peptidoglycan	29	polar body	93
peptidyl site	84	polar compound	15
perfusion	124	polar fiber	73
perikaryon	107	polar lobe	99
perinuclear space	36	polarization microscopy	124
perinucleolar chromatin	36	pollen grain	93
peripheral lymphoid tissue	119	pollen tube	93
peripheral membrane protein	18	pollination	93
peripheral nervous system	108	polycistronic mRNA	84
permease	23	polymer	11
peroxisome	29	polymorphonuclear leukocyte	37
pH	7	polymorphy	79
phagaocytic vesicle	37	polypeptide	18
phagocyte	37	polyribosome	84
phagocytosis	42	polysaccharide	11
phalloidin	114	polyspermy	93
phase-contrast microscopy	124	positional information	99
phenotype	78	positive regulation	48
phosphatidic acid	14	positive supercoil	63
phosphodiester bond	53	poststaining	124
phosphoester bond	23	postsynaptic	108
phosphogluconate pathway	54	postsynaptic membrane	108
phosphoglyceride	14	posttranscriptional control	84
phospholipid	14	posttranscriptional processing	84
phospholipid transfer protein	18	posttranslational control	84
phosphorylating transport	42	potential energy	7
phosphorylation	23	presnyaptic membrane	108
photoautotroph	103	presynaptic	108
photoauxotroph	103	Pribnow box	84
photochemical reduction	57	primary cell wall	37
photoexcitation	57	primary fixation	124
photoheterotroph	103	primary follicle	93
photolithotroph	103	primary germ layer	93
photolysis	57	primary immune response	120
photophosphorylation	57	primary lysosome	29
photoreduction	57	primary oocyte	93
photorespiration	54	primary spermatocyte	93
photosynthesis	57	primary structure	18
photosynthetic unit	58	primase	67
photosystem	58	primer	67
photosystem I	58	primer recognition factor	67
photosystem II	58	primitive atmosphere	101
phototroph	103	primordial germ cell	94
phragmoplast	73	primosome	67
phycobilin	58	process	108
physiology	3	progesterone	94
pigment	58	projector lens	124
pilus	89	prokaryotic	3

promoter	89	responsiveness	4
prophage	89	resting potential	108
prophase	73	restriction enzyme	64
prophase I	73	restriction point	73
prosthetic group	24	restriction site	64
protamine	94	retinene	108
protease	24	retrovirus	90
protein	18	reverse transcriptase	90
proteoliposome	108	reversible inhibitor	24
proteolysis	24	rhodopsin	108
proto-oncogene	79	ribonuclease	64
protofilament	29	ribonucleic acid (RNA)	64
proton	7	ribose	11
proton motive force	43	ribosomal RNA (rRNA)	64
protoplasm	3	ribosomal subunit	30
protozoan	89	ribosome	30
pseudoplasmodium	94	RNA polymerase	68
pseudopod	94	RNA splicing	84
purine	63	RNA-DNA hybridization	84
pyrimidine	63	rough endoplasmic reticulum	30
pyrimidine dimer	79	S phase	73
quantum requirement	58	sarcoma	79
quaternary structure	18	sarcomere	114
radial spoke	114	sarcoplasm	114
radiation	7	sarcoplasmic reticulum	114
random coil	18	saturated fatty acid	15
reactant	7	saturation	15
reaction center	58	scanning electron microscope	125
receptor	37	scanning transmission electron	
receptor-mediated endocytosis		microsc.	125
(RME)	43	Schwann cell	109
recessive	79	scientific method	4
recombinant DNA	67	second law of thermodynamics	48
recombinant DNA technology	79	second messenger	30
recombination	79	secondary cell wall	37
reduction	8	secondary electron	24
refractive index	124	secondary immune response	120
refractory period	108	secondary lysosome	30
regulator gene	89	secondary oocyte	94
regulatory T cell	120	secondary spermatocyte	94
reiterated sequence	63	secondary structure	18
relaxed state	63	secretion	4
remission	79	secretory component	120
renaturation	18	secretory granule	30
repair endonuclease	63	secretory protein	30
repair synthesis	63	sedimentation coefficient (S)	30
replication	67	selective permeability	43
replication bubble	67	self-assembly	30
replication fork	68	semiautonomous organelle	30
replicon	64	semiconservative replication	68
repression	90	seminiferous tubule	94
repressor	90	sensory neuron	109
reproduction	94	sex chromosome	33
residual body	37	sexual reproduction	94
resistance factor	90	shadowing	125
resistivity	4	shuttle streaming	114
resolution	125	sigma factor	84
respiration	54	simple active transport	43
respiratory assembly	54	sister chromatid	73
respiratory control	54	site-specific mutagenesis	79

sliding filament theory	114	synaptic knob	109
sliding microtubule model	115	synaptic vesicle	109
small ribosomal unit	30	synaptonemal complex	74
smooth endoplasmic reticulum	31	syncytium	115
sodium cotransport	43	synthetic work	24
sodium-potassium pump	43	system	54
sol	31	T lymphocyte	120
solute	8	TATA box	85
solution	8	tau protein	31
solvent	8	tautomerization	54
somatic cell	94	teleocentric chromosome	74
somatic nervous system	109	telophase	74
spacer DNA	85	telophase I	74
spacer sequence	64	temperate bacteriophage	90
spatial summation	109	template	85
specific heat	48	temporal summation	109
specificity	19	teratogen	79
spermatogenesis	95	terminal cell	95
spermatogonium	95	terminal glycosylation	19
spermatozoa	95	terminal oxidase	24
sphere of hydration	8	termination signal	85
sphingolipid	15	tertiary structure	19
sphingomyelin	15	testis	95
sphingosine	15	testosterone	95
spindle fiber	33	tetrad	74
spindle pole	73	tetrose	12
spontaneous	48	theory	4
spore	73	thermodynamics	49
sporic meiosis	73	thick filament	115
sporophyte	73	thin filament	115
spot desmosome	37	thin section	125
stage	125	threshold stimulus	109
staining	125	thylakoid	59
starch	12	thymine	64
steady state	48	tight junction	37
stem cell	99	tissue	4
stereo pair	125	tonicity	43
stereoisomer	12	tonofilament	37
steroid	15	topoisomerase	64
stimulus	109	totipotent	99
stoma	58	trace element	8
stop codon	85	transamination	19
storage macromolecule	12	transcellular transport	43
striated muscle	115	transcription	85
stroma lamella	31	transcription unit	85
structural gene	64	transcriptional control	85
structural macromolecule	31	transducin	110
substrate activation	24	transducing bacteriophage	90
substrate analog	24	transduction	110
substrate induction	85	transfer RNA	85
substrate-level phosphorylation	54	transformation	90
sucrose	12	translation	86
supercoil	64	translational control	86
suppressor strain	85	translocation	86
supramolecular structure	31	transmission electron microscope	
surface tension	43	(TEM)	125
surface-to-volume ratio	43	transplantation antigen	120
surroundings	48	transport vesicle	38
synapse	109	transverse (T) tubule	115
synaptic cleft	109	transverse diffusion	43

trigger protein	74	vacuum evaporator	126
triglyceride	15	valence	8
triose	12	variable domain	120
triple bond	24	vectorial pumping	44
triplet code	86	vesicle	38
tropomyosin	115	virulent bacteriophage	90
troponin	115	virus	90
true-breeding	79	vitelline membrane	95
tryptophan operon	90	wavelength	126
tubulin	31	weight	4
tumor necrosis factor	80	wild type	80
turnover	19	wobble	86
ultracentrifuge	125	work	49
ultramicrotome	125	zona pellucida	95
uncoupler	54	zygote	95
unit membrane	38	zygotene	74
unsaturated fatty acid	15	zymogen granule	31
uracil	65		
uridine triphosphate (UTP)	49		
vacuole	31		

www.ingramcontent.com/pod-product-compliance
Lightning Source LLC
Chambersburg PA
CBHW081129170526
45165CB00008B/2602